SOUL

OF A

LION

Zion as an adult

SOUL
OF A
LION

One Woman's Quest to Rescue
Africa's Wildlife Refugees

BARBARA BENNETT

Foreword by Marieta van der Merwe

■ NATIONAL GEOGRAPHIC

WASHINGTON, D.C.

For Carolyn Cooke 1949-2009

Published by the National Geographic Society
1145 17th Street N.W., Washington, D.C. 20036

ISBN: 978-1-4262-0654-2

Library of Congress Cataloging-in-Publication Data

Bennett, Barbara.
 Soul of a lion : one woman's quest to rescue Africa's wildlife refugees / Barbara Bennett ; foreword by Marieta
van der Merwe.
   p. cm.
 Includes index.
 ISBN 978-1-4262-0654-2 (hardcover) -- ISBN 978-1-4262-0667-2 (e-book)
 1. Van der Merwe, Marieta. 2. Wildlife conservationists--Namibia--Biography. 3. Wildlife rehabilitators-
-Namibia--Biography. 4. Harnas Wildlife Foundation. 5. Wildlife refuges--Namibia. 6. Wildlife rescue--
Namibia. 7. Wildlife conservation--Namibia. 8. Namibia--Environmental conditions. 9. Namibia--Biography.
I. Title.
 QL84.6.N3B46 2010
 333.95'416096881--dc22

                    2010015023

The National Geographic Society is one of the world's largest nonprofit scientific and educational
organizations. Founded in 1888 to "increase and diffuse geographic knowledge," the Society works to
inspire people to care about the planet. It reaches more than 325 million people worldwide each month
through its official journal, *National Geographic*, and other magazines; National Geographic Channel;
television documentaries; music; radio; films; books; DVDs; maps; exhibitions; school publishing
programs; interactive media; and merchandise. National Geographic has funded more than 9,000
scientific research, conservation and exploration projects and supports an education program combating
geographic illiteracy.

For more information, please call 1-800-NGS LINE (647-5463)
or write to the following address:

National Geographic Society
1145 17th Street N.W.
Washington, D.C. 20036-4688 U.S.A.

Visit us online at www.nationalgeographic.com

For information about special discounts for bulk purchases, please contact
National Geographic Books Special Sales: ngspecsales@ngs.org

For rights or permissions inquiries, please contact National Geographic Books
Subsidiary Rights: ngbookrights@ngs.org

Photo Credits: 2, Christophe LePetit; 26, Christophe LePetit; 50, van der Merwe family
photo; 74, van der Merwe family photo; 96, Christophe LePetit; 112, Christophe LePetit; 130,
Christophe LePetit; 140, van der Merwe family photo; 156, Barbara Bennett; 166, van der
Merwe family photo; 180, van der Merwe family photo; 192, Cornelia Achtel; 208, property of
Barbara Bennett; 240, Christophe LePetit; 254, Barbara Bennett; 268, Christophe LePetit; 280,
Jo van der Merwe

Designed by Al Morrow

Printed in the United States of America

10/WCPF=CML/1

CONTENTS

FOREWORD

by Marieta van der Merwe

I LIVE MY DREAM EVERY DAY.

One volunteer told me she wanted to be like me and asked me how she could do that, and I told her that it had to come from her heart; I couldn't tell her what to do because it had to come from inside her. I didn't start Harnas because I wanted to be famous, and if she wanted to be well known because of the animals, it just wouldn't work.

Taking care of animals was a choice for me, and when you make a choice like mine, you also make a choice not to be so involved in the world. You can't go to the cinema or out to dinner or to parties, because you might have a baby baboon with you, and they aren't allowed to be there. You must make a choice when you go on holiday, because you must pick places where your animals can go, too.

But for me it was an easy choice.

If an animal needs something, I know that I will do whatever I can to help. If a baby baboon needs a bottle in the middle of the night, I won't say, "I'll do that tomorrow." I do what the animals need, when they need it. If you don't believe the animal comes first, you will never be successful at this.

When I first started Harnas, I did all the work myself. I would get up at five o'clock in the morning and set out all the bottles that needed to be filled, all the medicine that needed to be given, all the meat that needed to be fed to the animals. I had a big table where I would put all of these things, and then I'd start with the babies that needed to be given bottles. Then I'd feed animals that needed meat. Then I'd give injections and pills that needed to be given. When I had finished, it would be time to prepare bottles again. And this would happen over and over the whole day, until it was night.

Life is easier now because I have so many volunteers and workers to help me. I also have more knowledge, knowledge that I got through my own reading, asking questions, and especially through experience. When I got my first volunteers, they started asking me questions, and when we received guests they also asked me questions, and I realized I had knowledge from working with animals every day. By watching animals, you learn how they act, what they do, and how they think.

Now other people call me about this animal or that animal. It's a good feeling when people ask me for the knowledge I've gained through the years, because I can pass on what I've learned.

Part of an old Chinese proverb states: "If you want to be happy for a lifetime, love your work." There is a lot of value in this for me. It has become part of my entire existence. Reading this book, you will become aware of the beauty of this day-to-day way of living, but you should also notice the reality of a dream sometimes broken. Financial shortages, the increasing challenges of more orphans and problem animals within a limited pool of resources, and the loss of a beloved animal are parts and pieces of this ever-changing environment.

Sometimes I wonder why God gave this for me to do. With millions of people to do this one thing, why did he choose me? Maybe he knows that if I do this—taking care of the animals—I will never stop doing it and won't do it halfway. I am the happiest woman on Earth, not only because I can do what I've always dreamed about but because I have the opportunity to share my dream with thousands of people in Namibia and around the world. I am privileged to be given the opportunity to play a vital role in taking care of God's creations. For me, nothing could be better than this, being a caretaker of animals.

PREFACE

*Animals are such agreeable friends—they ask no questions,
they pass no criticisms.*

—George Eliot

IT WAS A BEAUTIFUL AFRICA MORNING, AND I LAY IN MY
bed listening to the sounds of wild animals waking up outside—
lions roaring their territory, baboons warning of intruders, and
the general readjustment of life after a night's sleep. I heard what
I thought was a cat jumping from a tree onto my roof. "Heavy
cat!" I thought." Then I heard it: "Uunnhh, uunnhh, uunnhh,"
the unmistakable call of a baboon, followed by another thud on
the roof. I jumped out of bed and looked out the window.

About a dozen baboons were in my yard, having a blast. They
swung from the trees, climbed on my banister and picnic table,
chased each other around the yard, and examined every item in
the outside garbage can—now overturned. It held peanut shells,
orange peels, granola bar wrappers, an empty Diet Coke can, and

used Kleenexes. All these artifacts were spread out on the ground so they could be examined more closely.

Then I heard a thunderous "rrrreeeeeekkkkkk" and looked up to see my corrugated tin roof being ripped back by a good-sized male baboon, creating a triangular opening to the sky, about the size of a dinner plate. Suddenly the animal thrust its black, hairy head through the hole and gazed down at me—not threateningly but with curiosity. The intrusion was still unsettling, to say the least, and I grabbed the only weapon I had—my boot—and threw it at him, shouting "NO!" He let the roof go, and I heard his feet thudding overhead. Through my window I watched him jump to the ground. I stomped on the wooden floor and pounded on the wooden walls, making noise that scared away all of them but the big one who had peeled back my roof. Off the others went, looking for new adventures, a new party, new people to alarm, new garbage cans to explore.

With the one remaining baboon outside now sitting on my front porch, I knew I had to wait it out until someone found me—no phone in the cabin. Adult male baboons can be aggressive, weigh up to 90 pounds, and have the strength of seven men, so I wasn't willing to wrangle with him. I threw one of my remaining two oranges out the window as far as I could, hoping he would be distracted by the offering and go away, but he retrieved it, brought it back to the porch, and started peeling and eating it while balancing on my banister.

He settled in, peering at me through the window, looking remarkably human with his curious brown eyes. The hole he had made was still open to the sky, and I couldn't reach high enough to pull the corner of the roof back down. I feared he might decide to enlarge it enough to come inside, so I quickly got dressed.

I decided if people were going to find me, torn, bleeding, and ripped apart, I wasn't going to be wearing my polar bear flannel pajamas. Then we waited—and waited some more. I tried to find things to keep me busy. I made coffee, ate a granola bar, and changed the sheets on my bed. I took some pictures of him and even made a short video, stating my final wishes in case things didn't work out. We stared at each other through the window some more. Every time I'd walk somewhere in the cottage to do something other than gaze into his eyes, he'd lean out from his perch on the banister and watch my movements.

After an hour and forty-five minutes, the Bushmen arrived to chase him home. Later I learned that it was their fault that the baboons were out. They were to have filled in the baboons' waterhole but had forgotten. The animals were thirsty and looking for water. Human error almost always causes sprees like this, and you can't blame the animals.

I went to the office to report the incident, and as the adrenaline drained from my body, I started to shake. It had been an interesting but, in many ways, a very trying morning. I had been brave, even stoic, keeping my sense of humor about me, but on the way to the office, I realized how tense I had been. As I started to tell the story to the office manager, Melanie—I'll admit it—I fell apart. I didn't cry, but I did some hysterical raving, I was told later. Apparently, at some point I wailed, "I'm from the freakin' suburbs! What am I doing here?" Melanie slipped me one and then another cup of coffee liberally laced with whiskey, and I started feeling better. Soon I was laughing about the incident—before I passed out.

HOW I ENDED UP IN SUCH A STRANGE PREDICAMENT isn't so odd when you consider my childhood. Practically every photograph of me as a child shows me holding an animal of some kind—cats, dogs, hamsters, turtles. I simply loved animals, and this passion carried through to my adult years. Somehow, though, I ended up as an English professor at North Carolina State University, where the wildest thing I encountered was a rogue freshman enjoying freedom too much. Animals continued to fascinate and attract me, however, so when I earned a semester's sabbatical, I decided to do some volunteer work. My first thought was to do something wonderful, far away, with wild animals. I began to cast about for a unique opportunity. An article led me to a website describing the perfect situation: working with orphaned, injured, and abandoned wild animals in Namibia, Africa, at a place called Harnas Wildlife Sanctuary.

I had to look up Namibia on a map. There it was, on the west coast of southern Africa. All I had heard about it was what most people know: Namibia is the country where Angelina Jolie and Brad Pitt had gone to have their baby, Shiloh. I figured their decision to give birth in a faraway African nation was nothing more than a publicity stunt, not understanding at that time Jolie's deeper connection to Harnas, her dedication to conservation, and her patronage of the sanctuary. I didn't learn about that until much later, while I was actually visiting Harnas.

Though Namibia was an unfamiliar shape on a map and Harnas was only a dot on that shape, it took me less than a second to decide to go. I applied, was accepted, made plane reservations, and started filling an upstairs bedroom with "Africa stuff"—mosquito net, sleeping bag, water purifier, cargo pants. Everyone I knew began asking, "Aren't you scared to go alone? Aren't you afraid no one will pick you up at the airport? Aren't you afraid you'll get eaten by a lion? Aren't you afraid of getting a fatal disease?" Truthfully, I hadn't thought much about any of those possibilities (other than getting the right inoculations) till then. All I thought was that I'd be in close contact with wild animals. I didn't think beyond that exciting thrilling expectation.

In February 2007, I headed for Namibia on an overnight flight. In the morning, as we circled the capital city of Windhoek, I looked out the window at the miles of empty land around the mile-high city. I had learned a great deal about Namibia in the months before my trip. Cradled between Angola to the north, South Africa to the south, the Atlantic Ocean to the west, and the Kalahari Desert and Botswana to the east, Namibia is remarkable in its variety of landscapes—spectacular sand dunes sculpted by fierce desert winds, beach resorts resembling traditional German villages left behind by 19th-century colonial settlers, stark water holes in the middle of dry, grassy savannahs, and rivers passing through ancient canyons.

Entering customs, I was glad—and a bit relieved—to see that someone was waiting for me, holding a sign with my name on it. The man, who spoke only limited English, carried my larger bag, even though he was much shorter and lighter than I. We walked out of the airport into brilliant sunshine blazing from a Wedgwood-blue sky. My pale winter skin prickled in the African summer and began to redden almost immediately. So I put on my hat until we got to the Jeep and joined four other volunteers who had arrived the day before and stayed the night at a youth hostel. We began our journey into the bush—stopping at various places to buy gasoline, food, and boxes of indistinguishable supplies.

With all the stops, what might have been a three-hour trip turned into a five-hour journey. We passed from the high plateaus of Windhoek, with its rolling hills and landscape of low trees and bushes east toward Gobabis. We descended gradually and the trees became fewer, shorter, and lighter in color. The land flattened out for long stretches of grasslands, dotted with clusters of thorny bushes or an occasional acacia tree. We turned north after a stop at Gobabis for food and drinks at a simple cement-block store with few choices and no frills. America, with its endless variety of foods and alluring facades was long gone.

The sun set abruptly. When we arrived at Harnas it was dark, and I couldn't see beyond the house where I was assigned to stay. My three roommates introduced themselves, but because I had been traveling for over 30 hours, sleep was the only thing on my mind. Once I was in bed and drifting off, I wondered why children were screaming at each other outside my window. Not until later did I realize I had been listening to baboons having their usual nighttime disputes.

The next morning I woke up before my roomies, pulled on some clothes, and walked outside to a large lawn area, crisscrossed with rock-lined paths winding through gardens with palm trees, cactus, and small ponds. Was I dreaming? Ostriches were parading around, as were small antelope-type animals, huge tortoises, a goat or two, some dogs and cats, and a few volunteers carrying baby baboons—in diapers! What kind of place was this? And what kind of people came here?

People like me, it turns out: people who would rather spend their holidays dirty, smelly, and covered in urine, blood, and sweat than lounge at the swankiest hotel in Paris. I was among my own species, so to speak. We all understood each other because we all loved animals. My roommates were tall, attractive young women from England, Germany, and the Netherlands. Most people would expect to see such beauties draped in designer fashions, sitting at an outdoor café, sipping fine wine. But they went without makeup, cut up fresh donkey meat for the carnivores, and slept with baby animals. Outside the gates of Harnas we might have been regarded as quirky because of our choice for a holiday excursion, but inside Harnas, we were considered hardworking, compassionate, dedicated volunteers.

The work was unending but joyful. One of my assigned chores was making sure the four baby baboons got playtime on the lawn each afternoon. I'd go to their enclosure and open the gate. Out they would scamper, the three smallest—Moses, Ita, and Ovambo—jumping into my arms, and the largest one, Jacob, hooking his limbs around one of my legs. Swinging my leg stiffly to give Jacob a ride, I walked through the gate and out to the lawn. That time of day, volunteers were usually sitting under a tree to escape the midday African heat. As soon as the baboons saw them, they'd run across the lawn and pounce on the volunteers; they'd then climb the tree and jump down into the arms of the volunteers. The babies would spend the next hour or so checking people's pockets, unbuttoning shirts, detaching and attaching anything Velcro, and stealing sips of Coke and Fanta.

Some evenings we went on walks with the half-grown lion cubs, Zion and Trust, who amused themselves by turning and jumping on us in pretend-pounce. We followed all the rules—don't scream or run, keep them in front of you, stay standing up—but they still found a way to surprise-attack. Nothing compares to having a 250-pound lion run right at you, jump up, wrap his front paws around your waist, and play-bite your legs and arms. Every scrape was a badge of honor. We all hoped one of these injuries would scar. Back at home, the account of the injury would be the cocktail party story that topped all others.

Another assignment was to feed Boertjie, a large baboon with epilepsy and Down syndrome. One evening I took his bucket of fruit slices out to his enclosure, an area along one length of a four-sided dirt lot. His enclosure shared one side with Gumbi, the brown hyena. Another side had a gate leading to the road heading toward the outer enclosures, another gate opened onto the lawn area. The fourth side, directly behind me as I sat down to feed Boertjie through his fence, bordered on the truck garages and the enclosure holding Zion and Trust.

I sat on the ground, cross-legged, and started to talk to Boertjie, taking breaks to slip pieces of apples, oranges, bananas, or grapes through the fence. When he opened his mouth to take the fruit, he exposed his massive canines, two inches long and powerful enough to crush the spine of an antelope. He'd always wait, though, until my hand was back through the bars before he closed his mouth and began to chew. He chewed very slowly, enjoying the human contact and the attention he craved. The only thing different I noticed that night was that his eyes seemed to wander. He usually locked his gaze onto mine, as if absorbing my words. But I chalked up Boertjie's distraction to his ailments and focused on keeping up my lively chatter.

After 30 minutes, I gave Boertjie his last piece of apple, stood, brushed off my shorts, and turned around. I inhaled sharply in alarm. Not five yards from me were Zion and Trust, lying in wait and watching me with great interest. I thought, "How did they get in here!" A surge of adrenaline feels much like an electric shock surging through your body; I froze. All I could do was widen my gaze and take deep breaths. And try to think.

A Bushman, I learned later, had left the gate unlocked. Hearing my voice as I talked to Boertjie, the two lions had come over to see what was going on, and whether they could join in the fun. I had already broken two major rules without knowing it: I had made myself small by sitting down, and I had turned my back on two lions. I tried to keep from breaking the third: Don't scream and run.

After the initial shock had diminished, I slowly let out a deep breath and tried to do what I had been taught. I spoke authoritatively to the lions and began walking toward them. "Come come come!" I said, shooing my arms as I approached them. Like children who have been told to come in from playing on a fine summer evening, the big cats reluctantly stood up and started walking toward their gate. I followed them, continuing my verbal encouragement and gently prodding their hindquarters with small pats. Like obedient youngsters, they sauntered through the gate and headed for their favorite sleeping platform under a large tree. I closed and locked the gate, shaking my head and wondering who would believe this.

I turned around and looked at Boertjie, trying to imagine the whole escapade from his perspective. He'd been eating his evening meal and listening to me chatter. Suddenly, he sees two lions padding up behind me. "Ah!" he thinks. "Dinner and a show!" Perhaps the lions were taken with the cadence of my voice. Perhaps they recognized me from our playtimes together; whatever went through their lion minds, they decided to lie down and listen to me rather than pounce on me. During the meal, Boertjie had kept an eye on the beasts beyond my shoulders, clearly feeling safe himself but perhaps speculating about how long this peaceful coexistence would last. He must have been quite amused— and maybe a bit disappointed—when the encounter ended so anticlimactically, at least for him.

During my two-week sojourn at Harnas, I worked harder than I had ever worked in my life, and I got dirtier and more scratched and bruised than I had thought possible. And I adored every second of it. I got to know the van der Merwe family a little, but I could tell they kept close company. I was mystified and awestruck by Marieta van der Merwe, the woman who had started the sanctuary, who could identify almost a hundred baboons by face and name, and who seemed to fear nothing.

My visit went by like fire through the savannah, and when it was time to go home, I cried all the way to the airport. I kept apologizing to the driver until he said, "Everybody cries when they leave." Because he saw animal lovers like me every day, he understood completely. When I returned to North Carolina, though, I found that people expected me to show my pictures and then get back to "real life." I became depressed. I had experienced such an emotional high the past few weeks—an intensely rewarding experience—that the drop-off afterward was severe.

Soon after, in the middle of the night, I woke with the clear certainty that I wasn't finished with Harnas. I needed to write about the place—about Marieta, about the selfless volunteers, and about the healing of both animals and people that goes on there. Along the way, the book could teach people about the importance of conservation. Writing it, I could combine the things I love: books, animals, teaching, and now Harnas. I contacted Marieta and her family, and they agreed to share their lives and work with me.

In June 2008 and again in December 2008, I returned to Harnas for a month. I went with a clear idea of the book I could write—a linear story about Marieta and the evolution of Harnas. After a few days, though, I felt the idea dissolve, to be replaced by a much more complex and interesting story of the people and animals on the farm. The story became a web, with Marieta at the center. Strands headed out in all directions, connecting and reconnecting in unexpected places. The story started to be more about the cycle of healing than merely the creation of an animal refuge. Animals come to Harnas to be healed—by people who are then healed through their service to the animals.

My subsequent experiences at Harnas were different. I was different and my assignment was different. Animals still played a major part in my daily life on the farm, but now I saw that the stories of the animals were intertwined with the stories of the people. My best thinking came while I sat in the cheetah enclosure, brushing burrs out of the animals' fur, or sitting in front of the fence, watching baboons chase each other, jumping, and swinging like acrobats, or drinking a cup of coffee in the morning, conversing through the fence with Elsa and Sarah—two large, tawny lionesses.

Marieta and the van der Merwe family, the office workers, the Bushmen and the local Africans who worked on the farm, the volunteers, and the tourists—all of our lives were affected by the lives of the animals. Once I became a part of the web of connection at Harnas, I moved around the farm with a new reverence for Marieta, her family, and what they were accomplishing.

My sojourns weren't always easy, as living on a wildlife farm can be chaotic at times, and adventures happen every day. One morning I was out taking pictures at the lion enclosure near my bungalow when I noticed some baboons running along the road, rather than behind the electric fence where they were supposed to be. I walked back to the main office and reported what I'd seen. The electricity that runs along those fences had been erratic, and baboons—unlike lions, who check the electricity once and then assume it's always on—take turns checking it several times a day. When it's off, they go over the top like convicts in a massive prison break.

At the office, Jonette, asked me, "Is your house secure?"

"Yes, it's locked."

"Windows, too?"

"Uh, no. I left my windows open to catch the breeze."

She winced, and I frowned in dismay. Out the door and back to my house I ran, arriving just in time to see Bushmen using rakes to chase baboons in and around my house. The monkeys were spilling out of my windows like tipplers escaping the police in a raid at a speakeasy.

Once the baboons were out of sight, Bushmen still chasing and shouting at them, I looked in my window to make sure no creatures were still inside pillaging. I unlocked the door and entered cautiously. The baboons had clearly partied in my house. As I walked in, the debris crunched under my shoes; they had eaten all my peanuts, saved to give to the baby monkeys, and left behind the shells. They'd opened a bag of Cheetos and thrown them up in the air like confetti. They'd opened my container of instant coffee and scattered it on the clean, folded clothes lying on my bed. They'd stolen most of my oranges but left me the skins. Fortunately, the door to my bathroom had been closed. I laughed, then I wailed, then I laughed again. And then I cleaned it up. I had been initiated by the baboons, learning their stringent discipline with every peanut shell and orange Cheeto heaped in my dustpan.

Experiences like the one I shared with the baboons filled my days when I wasn't listening to stories and researching facts. As

I learned the details of Marieta's life and how much she and her family had accomplished and overcome in creating the preserve, I became an even bigger fan. In the evenings, she and I took walks through the bush while I recorded her stories. Her English was halting at times, Afrikaans being her first language, and she spoke only in the present tense. Sometimes her daughters-in-law Jo and Melanie came with us to help her find the right words when she struggled. I marveled at her tenacity and her casual bravery. One evening as the sun was setting in an orange sky, we walked along a dirt path next to the lions' enclosures while she talked and I gazed into the darkness, hoping to catch sight of a pair of yellow eyes through the fence. Without warning she threw her arm across my chest, jarring me to a standstill.

"Snaaaaake!"

I glanced down and saw a curled-up puff adder—one of the deadliest snakes in Namibia—just six inches in front of my boot. We backed away cautiously and circled around the spot before continuing. I was speechless for ten minutes, shocked at the closeness of disaster and death, but Marieta talked on as if nothing had happened. Just another animal encounter in the wilds of Africa.

I have more personal stories to tell, but this book isn't really about me. It's about Marieta and Harnas—and I have plenty of amazing and entertaining stories collected there. But perhaps you'll permit me one last personal story—the one that means the most to me. The one that changed how I feel about the rest of my life.

One night near the end of my second visit, I slept out in one of the cheetah enclosures by myself with two female cheetahs, Cleopatra and Pride. These two beauties were kittens during my first visit, and I had fed them, petted them, brushed them, and planted kisses on their sweet faces. Both were born in 2006 on the farm, but originally Pride grew up with Trust, the lion cub. Eventually, though, Trust became too powerful for her, and so she was paired with Cleopatra. When Cleo and Pride were playing one day as cubs, Cleo ran headlong into a pole and suffered some brain damage. She walks a little unsteadily, but otherwise is fine—and is sweet and gentle. They are now a tightly knit pair, and I had grown to adore them.

Volunteers sleep out with animals sometimes, usually in pairs or groups, but I never had, so it was a significant night for me. Originally, one of the women from the office was going to go with me, but she backed out at the last minute. I had been planning on it for days, though, so I went alone.

I dragged a foam mattress out to the cheetahs' enclosure, laid my sleeping bag on it, and went to find the cats to alert them to my presence—no surprises, please. Then I snuggled down into my bag. About 15 minutes later, Pride glided over. I heard her purr long before I saw those amber eyes emerge out of the African night. Cheetahs have the loudest purr of all cats, and Pride's is especially strong. She strolled over to me, put one paw on each side of my face, tickled my cheek with her tongue, and plopped down with her neck stretched across mine. Then she really let that purr roll forth. I felt as if I were inside one of those vibrating beds that take quarters in motel rooms. I wrapped my arms around her in a furry embrace, and in a few minutes we both fell asleep. After an hour, she stood up, stretched, and wandered away.

Sometime later, she came back. Then Cleo joined us—stretching her body across my legs, resting her head on my lap. The three of us spent the rest of the night like a litter of kittens all tangled up together. It was a cold night, but I had almost 300 pounds of cheetah keeping me warm. Burying my nose in Pride's neck, my hands around her chest, I stayed toasty. Once I woke up to find I had been displaced and was sleeping on hard ground, my mattress occupied by both cats. Cleo was splayed on her back with paws in the air and white belly exposed trustingly. I scrambled up as best I could in my sleeping bag and "scootched" back over, pushing my way in between the cheetahs to reclaim my spot on the mattress.

Later during the night, I awoke and looked up at the moon and the unfamiliar stars in the African sky. I had a cheetah's head buried in my neck and another purring in my lap, and I thought, "If nothing good ever happens to me again, it's all right. I've had this—and this is worth more than anyone deserves."

The sun came up about 5:30. We snuggled some more, and they groomed each other—and me, licking my face and arms till they were raw. At seven o'clock I said my good-byes and went home to a hot shower, forever changed.

It's hard not to overuse the word "magical" when talking about scenes like this, Harnas, Marieta van der Merwe, and the animals she has rescued, but few situations deserve the word more. While each person who visits Harnas has a different experience, most come away with new reverence and love for the natural world that surrounds us but is too quickly eroding. Not everyone can go to Africa and see these disappearing wonders for themselves, but maybe hearing about Harnas will begin for you the same kind of transformation that happens to visitors. Whether in body or through story, it's a journey worth taking.

Marieta with her granddaughter Aviel and baby baboon Grace

chapter one

EXPECT *the* UNEXPECTED

In everyone's life, at some time, our inner fire goes out. It is then burst into flame by an encounter with another human being. We should all be thankful for those people who rekindle the inner spirit.

—Albert Schweitzer

THE AFRICAN SUN IS BEGINNING ITS ASCENT IN THE eastern sky. Just a hint of light is followed by streaks of peach, gold, and even purple. The chill of the desert night begins to disperse, and before long the land will be blazing hot.

From nearby, the crowing of roosters announces the day, but they seem to be confused, as they have been announcing the coming of day throughout the night. The more accurate proclamation comes from a different direction: a coughing sound, like a large man clearing his throat, but amplified. This coughing is answered with a similar guttural sound, distant but distinct, deep and almost godlike, filling the air with its insistence for attention. The grunt reverberates from another direction, and suddenly the volume and intensity rise to that primeval demand that has made

humans and animals alike fight the conflicting urges to hide and
to run to save themselves.

It is the lion's roar, answered first from one direction and then
another, increasing in degree until you can feel the vibrations in
your chest, each roar beating to the rhythm of your heart. When
the lions' calls have reached their peak, they begin to wane, until
the sounds are mere growls expelled with each deep animal
breath. In and out, in and out, the growls lessen, and then there
is stillness. Territory has been marked for the moment—as it will
be again at sunset and at tomorrow's sunrise and each day after.
So the day has begun at the Harnas Wildlife Sanctuary.

In her bedroom, Marieta van der Merwe is beginning to
stretch, having heard the lion alarm clock. As she sits up in bed,
she takes stock of the babies surrounding her. A three-month-
old baboon, Ita, is strapped to Marieta's waist with a scarf, sucking on her spindly thumb, eyelids bluish and still closed. Ita was
brought to Harnas because her mother had been shot by a trophy
hunter who then found the baby clinging to her dead mother's
underside. The hunter brought the baby baboon, in shock and
dehydrated, to Harnas rather than selling her, and Marieta took
her willingly, even gratefully, cradling the small, fuzzy head the
size of a baseball, gazing directly into her eyes, and making the
small and fast tongue movements in and out of her mouth—
exactly what the baby's biological mother would do—signifying
love, safety, and connection. Ita found a new mother, and she
clings to Marieta obsessively, gradually recovering from the shock
of her mother's deathly stillness next to her own beating heart.

Marieta unties and then rewraps Ita in a protective circle of
pillows, hearing only a small squeak in protest. She lays her next
to two other baby baboons, slightly older, who had been clinging

to each other in sleep. Both were brought to Harnas when their families were scared away from a water hole by tourists. The baboon mothers had to leave so quickly that they weren't able to gather up their brave, exploring infants. The tourists found the babies and asked around about who might care for them. "Harnas" was what the locals told them over and over. Concerned ecotourists, the people drove hours out of their way to bring the babies to Harnas, where they were taken in with a smile. Marieta never says no to animals. It is a rule she lives by.

Other animal noises begin to attract Marieta's attention, and she reaches down to a basket next to her bed and pulls out a baby leopard named Lost and two Chihuahuas named Honey and Perky, whose job it is to comfort the leopard cub during the night, giving her heat and reassurance. These dogs have a long history of nurturing all kinds of orphaned animals. Honey, for example, nursed a young leopard named Missy Jo along with her own puppies (supplemented with extra bottles for the quickly growing feline). Honey treated Missy Jo like one of her own, grooming her, comforting her, and giving her milk. Honey's newest orphan, Lost, sometimes sucks on the little dog's ear. As any good mother would, Honey bows to the cub's will.

Marieta gently lifts the two-month-old leopard onto her lap. Lost's mother, too, was killed by a hunter. She wandered, nearly dead from dehydration and starvation, until a farmer's wife found her outside the kitchen door, mewing for help. A leopard is nocturnal, so Lost has been moving around throughout the night. As always, she is ready to have fun. Her bright blue eyes sparkle with the prospect of playing pounce. Fond of the game herself, Marieta moves her hand underneath the covers, first slowly and then quickly, attracting the attention of the stealthiest hunter in the African wild. The cub responds with a gleeful leap through the air, landing on the shifting blanket, with her teeth snapping and her claws exposed.

Marieta hoots with laughter. She is not the sort of woman who hides behind her hand and laughs discreetly. She throws her head back and enjoys every chortle and guffaw that life has to offer, especially when she is amused by one of her nearly four hundred rescued animals. She finds delight in the faces, movements, personalities, and growth of each creature, and she responds to them with love and laughter. In return, they offer her loyalty, endless love, and frequent entertainment. Even the name Harnas (in modern English spelled "harness"), prophetically given to the farm before the van der Merwe family ever lived there, symbolizes the mission of safeguarding wildlife: "harnas" was the protective armor used by medieval knights.

Generally, five kinds of animals end up at Harnas. First are the so-called problem animals, having become a problem mainly because of the loss of habitat caused by human expansion and the presence of livestock—cattle, goats, sheep, and chickens—easy prey found on local farms. (Why would a leopard spend time and energy chasing down a springbok, for example, when a passive calf is waiting in a nearby field?) Cheetahs, baboons, leopards, and a variety of smaller carnivores are included in this "problem" group. Lions used to be a problem but now are virtually extinct in wild areas of southern Africa. They were shot or poisoned whenever farmers discovered their presence near livestock.

Some wild predators can be relocated to national parks, sanctuaries, and wildlife farms, but these places simply do not have enough room for an expanding population. Plus, the official stance of the Namibian government can be summed up in simple terms: "If it pays, it stays." For instance, cheetahs that live in national parks—where tourists come to see them and stay in expensive lodges and eat in local restaurants—are worth protecting because they support the tourist industry. A cheetah in a remote area that might prey on goats or other livestock, on the other hand, makes no money for anyone and only hurts the local economy. Therefore, the government allows these animals to be shot without consequence.

Second, injured or imperfect animals arrive after people find them and either bring them directly to Harnas or to a vet. Sometimes the animal has been caught in a trap or hit by a car. Marieta pays for their medical care and gives them a place to live if they cannot be returned to the wild.

"Patcha," Marieta recalls, "is a leopard who lived on a farm near Harnas. When he started acting strangely—crazy with dangerous mood swings—he was taken to a clinic where he was diagnosed with hydrocephaly, or water on the brain. I refused to let the vet put him down, and brought him to Harnas instead. I gave him his own enclosure to live out his life peacefully where he wouldn't hurt others and others wouldn't hurt him.

"Another time, a cheetah named Bagheera was brought to Harnas with his two sisters when they were only two days old. I noticed right away that Bagheera had a problem because he wouldn't walk. He just dragged himself around by his front legs. *Och*, it broke my heart! The vet told me that his spine had been broken, and his back legs were paralyzed. Again, he told me to put Bagheera down, but I said no. We had a visitor then at Harnas who was very good with mechanical things, and he built Bagheera a wheelchair to hold his back legs so he could run. For the first time in his life, Bagheera ran like a cheetah should.

He didn't live long, but at least he lived long enough to be a real cheetah for a while."

A sad third portion of the population at Harnas consists of former pets. Some wild animals, so adorable as babies, grow up fast and become the large, sometimes aggressive animals they are meant to be. Fearful of their former sweet pets and unable to handle what they have become, people often resort to abusive treatment. Once, Marieta got a call to come get a baby baboon that people in Windhoek didn't want anymore.

"When my husband, Nick, and I arrived at their house, we found a young baboon covered in cigarette burns—on his face, his hands, every sensitive part of his sweet little body. The woman had bought the baby as a pet and named him Fransie, but her boyfriend didn't like the baboon. I think he was jealous of the attention his girlfriend was giving the baby and so he was burning him.

"We asked the boyfriend why he would do such a thing, and he said he was trying to teach the baboon to behave. If Fransie did something the man didn't like, he burned him as a way to 'teach' him discipline. But he didn't understand that baboon babies are much like human babies! They are always reaching for things and curious about every item around them, playing with things to see how they work. Fransie was just being a normal baby baboon and was being punished severely and unfairly.

"We took the baboon immediately to the vet, who was simply horrified that someone could do such a thing to a baby animal! By this time, the vet didn't even try to get us to put the baby down. He just gave us salve for the burns and pills to keep away infection. Once Fransie was healthy enough to live with the other baboons, he was fine—so happy to be allowed to act like a baboon needs to act and especially to be with other baboons. People don't realize that animals love to be with their own kind! They need that. And we were amazed that Fransie recovered completely and lived a long life at Harnas. Eventually, even his scars disappeared." Marieta laughs again and shakes her head. "Animals are so . . ."—she pauses as she searches for the word—". . . resilient!"

Another pet owner bought a baby baboon and named him Jacob, but she found that it was too much of a bother to cut a hole in Jacob's diaper to thread his tail through each time she put a new diaper on him. She took him to a vet and had his tail amputated—even though the tail is like a fifth limb for baboons, helping them move among branches and maintain their balance. Not long after having Jacob's tail amputated, the woman got tired of him and turned him over to Harnas. Having been raised as pets in houses with people, these animals are unable to return to the wild. Virtually all of them living at Harnas would be dead if not for Marieta.

Fourth, Namibia is full of orphaned animals, and Harnas is one of the few officially recognized orphanages for wild animals. Because hunters and poachers are always prowling for trophies, often killing an adult female and only then discovering her babies, Marieta's house has never lacked for abandoned infants. Since they are cared for by humans, these orphans are unable to return to the wild, and in truth, Namibia has lost most of its wild land except in the national parks. The country has been divided into farms, restricted for diamond and other gem mining, and developed as part of cities and towns. In the few "open" landscapes—such as the Herero tribal land—the game has been depleted almost completely through hunting, giving the few large predators left the choice of snatching livestock or dying.

Whiskey and Vodka, two caracals—similar to the lynx of North America—came from a farmer who had shot their mother. When he found the cubs, he tried to take care of them himself, but he really didn't know how, so eventually he brought them to Harnas. By then they were too old to be tamed and have lived their lives understandably distrustful of humans.

Other stories have better endings: four sister cheetah cubs—Leuki, Jeanie, Shingela, and Nikita—came to Harnas after their mother was shot by a farmer. These tiny cubs had severe calcium deficiency, but they grew out of that with the proper nutrition, and became healthy, trusting cheetahs. They love to be groomed by volunteers, and—unusual for cheetahs in captivity—Nikita has had two litters of cubs with a male from another group.

Fifth, and finally, baby animals are born on Harnas. Marieta's staff tries to keep animals that are not endangered from breeding by employing one form of birth control or another, but nature often complicates their efforts. As time passes, the number of breeding animals is being reduced as money for sterilization and the understanding of birth control in various species has increased. Male baboons can easily be castrated, for example, but if the female baboon goes into heat, wild male baboons will come from far and wide and somehow find a way over the electric fences.

"So now we sterilize the females instead of castrating the males," Marieta says while throwing up her hands. "This is more expensive and harder to do, of course, but in the end we will have more stability with our baboon numbers. It will take some time, though, and meanwhile, we see new babies riding on mothers' backs." In spite of her dilemma, Marieta smiles at the image this brings to mind.

Lions, because they are endangered, are carefully bred at Harnas. Elsa, Sian, Dewi, Kublai, Sher Kahn, Macho, Lerato, Zion, and Trust are just a handful of the lions born on Harnas. In addition, the African Wild Dog Project at Harnas was established to increase the wild dog population in Africa, as the animal is highly endangered. So far, this program has been successful, with puppies born on a regular basis. Eventually, the goal is to release selected captive dogs into a 25,000-acre reserve.

With every day so busy, Marieta doesn't spend much time playing with the animals in her bed. She gets up, stretches, and pulls open her gauze curtains to reveal another impossibly beautiful morning. Picking up Lost as the mother would, by the scruff of her neck, Marieta purrs to her as she takes her into the kitchen and places her on one of the two blanket-covered mattresses set on the floor each night. Eleven dogs of various sizes and breeds sleep here, usually curled up together in an endless tangle of yellows, browns, blacks, and whites. They are often joined by lion cubs, cheetah cubs, leopard cubs, baboon babies, caracals—and even once in a while, a warthog baby or a porcupine kitten. As youngsters, these animals don't seem to comprehend the strangeness of their multispecies litter. An artificial and temporary family they are, but one that works. Lost nudges aside the sleeping dogs with her nose, planting herself among them. The dogs part congenially, barely sighing in their sleep. In a matter of moments, all creatures are asleep again.

After she takes a few moments to herself in the bathroom, taking a quick bath, combing her curly blond hair, and swiftly glancing at her image in the mirror—possibly the last solitude she will enjoy all day—Marieta returns to the kitchen and begins filling various-sized bottles with the particular mixture of milk and vitamins each species needs.

A cacophony of sounds comes from the hallway nearby— meows and clicks and barks and human words as well as phrases like "lapa lapa lapa," "Mara," and unintelligible words from several languages. It is Tumela, Marieta's African Grey parrot—a birthday gift from her children who joke that she needed yet another animal to care for. Tumela has learned to mimic many kinds of animal noises at Harnas, as well as words in Afrikaans, English, Bushman, and various tribal languages. Soon, Bushwomen will move his cage outside to the courtyard where he can enjoy the sunshine, meow at the confused cats, and startle volunteers by calling to them in English as they walk by.

"Kasoepie!" Marieta yells, and a tiny Bushwoman appears from the stainless-steel kitchen where she and the other Bushwomen work. Their room is attached to Marieta's yellow-and-white kitchen, where everyone loves to gather. Kasoepie is dressed in layers of brightly colored skirts and aprons, with a scarf tied around her head in the traditional Bush style, wrapped and then tucked. At barely over four feet, Kasoepie is the size of a child, but her face shows her to be a woman who could be anywhere between 35 and 60. She will take orders from no one but Marieta. She completely ignores what others say and goes about her own business. Her light caramel-colored skin is stretched tight over high cheekbones surmounted by Asian-looking eyes. There is no confusing Black Africans with Bushmen. Their looks, physical size, coloring, and language are distinctly different. Kasoepie has been with Marieta for as long as they have been living on Harnas, and she is the only non–family member who gets away with telling Marieta, "Go to hell." Their arguments are legendary.

When Marieta and Nick were first building their house in the early 1970s, a Bushman tried to sell them two of his nine children, Kasoepie and her brother. The young couple refused, saying they would take the children in and care for them, but that they felt that people shouldn't be bought and sold! But the man persisted, saying he needed the money. In those days, some Africans paid Bushmen for their work with cups of alcohol, and as a result, this man was drunk most of the time. Finally, to save the children, Marieta and Nick paid the man and took the children in as workers. The brother eventually left Harnas, but Kasoepie stayed and helped raise all three of Marieta's children, becoming a part of the family. This particular morning, Marieta uses a combination of English, Afrikaans, and Nama words to instruct Kasoepie about various chores that need to be done, and Kasoepie disappears quietly to begin her work.

Jonette, a young and dark-haired member of the staff, arrives and makes coffee on her way to work in the main office. Nica, a yawning, slow-waking five-year-old with her grandmother's large blue eyes, appears from her mother Melanie's room and climbs onto one of the couches covered with blankets. These coverings hide the ravages of years of lion, cheetah, and leopard claws, not to mention the destruction wrought by dogs, cats, and children. Underneath the colorful blankets, the armrests show gouges deep into the foam, and the cushions are bumpy where pieces have been extracted. Nica dozes while a Sponge Bob cartoon from another world plays on the television.

Marieta casually calls to a small silver creature on the kitchen table and stretches out her arm: "Come, come come!" A small vervet monkey jumps gracefully from the tabletop to her shoulder. Once she passes close to the window, the monkey jumps to the windowsill, where he can watch the comings and goings of people and animals alike.

Another worker, Nicola, stands at the door leading to the food-prep area. "I need the keys to the freezer room," she announces. "The volunteers are waiting."

And Jonette, patiently waiting her turn, reminds Marieta that she needs the receipts from the day before. Being an office worker at Harnas is not like working in an office anywhere else. The job involves the usual routines of making coffee, talking on the telephone, and answering guests' questions, but any member of the staff might also be called upon to chase baboons back into their enclosure, drive to Gobabis to pick up boxes of intestines donated as animal food by a meat company, give a tour to a group of tourists who arrive unexpectedly and want to see and touch a cheetah, or hold down an injured meerkat while a co-worker cleans and bandages a wound; the vet is called in for only the most serious cases. Using the fax machine and filing? They're only a small part of the job.

Frikkie von Solms, Marieta's cousin, heads up the volunteer program. He enters the kitchen and begins fixing himself a cup of coffee. Frikkie is in his early sixties, but a life spent in the bush shows on him. He is as weathered as a piece of jerky, and he carries not an extra ounce on his skin-and-bones frame. His skin is tanned to a deep chestnut and his hair and mustache are sprinkled with gray. Although he seems to subsist on ciga-rettes, coffee, and whiskey, no one can outwalk Frikkie in the bush. Much younger volunteers and guests have to jog to keep up with him, and the only time he stops is when he wants to teach something to the breathless youngsters—like pointing out a poisonous tree, explaining how to tell direction when you are lost, or showing them a Bushman's deserted stick-and-grass shelter.

At 7:15, Marieta's son Schalk wanders in for some coffee and to gather information about the day's activities. Everyone who describes Schalk inevitably uses the word "Tarzan." A former World Cup rugby player, he is broad-shouldered and muscular. His hair is dark blond, wavy, and falls almost to his shoulders. He does not wear shoes—even when he walks through the thorny savannah. He is handsome, with a strong jaw, brown eyes that turn amber in the sun, and a movie-star smile.

Schalk is reluctant to talk, though, relating to animals more easily, and when he is forced to give interviews, it is clear he would rather be conversing with his big cats than talking to reporters with prying questions.

"I'm more like a leopard," he admits, "shy, a loner."

In this way, he is definitely his mother's child. Marieta claims that "being an only child made me suspicious of people. Animals were my first friends. My husband, Nick, was also an only child, and we shared this trait. I guess we've passed this on to Schalk as well as our daughter Marlice."

Of her three children, Schalk was the one Marieta worried about the least growing up on a wildlife farm. "I always knew he could take care of himself physically—even when he was as young as five years old. He would go out all day in the bush with his Bushman friend Gou, but I knew they'd be home that night. He was always strong, always independent, always careful."

Marieta knows that Schalk is less able to defend himself in verbal jousting. He doesn't join fast-paced conversations, and sometimes other people answer for him—which frustrates him. "I prefer to listen—or at least pretend to," he says with a wry smile. From the time he was a child, he has tended to stutter when he gets frustrated, so Marieta (or Schalk's wife, Jo) sometimes jumps in and takes his side. Always one to fight for the underdog, Marieta is fierce when her children are put in a situation that exposes their weaknesses.

On this particular morning, Schalk is concerned about a young baboon, Sarah, who has given birth overnight. Sarah has not been sterilized because she is still very young. A wild baboon jumped over the fence when she went into heat the first time and got her pregnant.

He explains the present situation to his mother. "Now that the baby has been born, Sarah doesn't seem to know what to do with it. It's still attached by the umbilical cord, and she's sitting on the cord. I can't even tell if the baby is all right. I think we should take it." He goes on, "She's too young, doesn't know how to care for it, might kill it. I think we should dart her, take the baby, and hand-raise it. Then we can sterilize both the mother and child."

"Let's wait another hour," Marieta counters. "If she still hasn't done something more, we'll go in."

"I still need the key!" Nicola implores from the kitchen door.

"Kasoepie! Where are my keys?" Marieta glances at a board on the wall, with hooks and a dozen sets of keys, but none is the one Nicola wants.

"Wait! Here they are," offers Melanie. Melanie, who is married to Marieta's older son Nico, is the photo negative of most of the family. While they are blond and curly-haired, she has straight dark hair, and while their eyes are light, hers are a chocolate brown. "They're behind the television. They must have fallen off the board."

At 7:45 the family gathers for the daily meeting. Marieta walks quickly across the courtyard to Schalk and Jo's house, newly renovated to create enough space for their two children: Samar, a five-year-old miniature of his father, and Aviel, a three-year-old who is embracing the princess within her (Disney reaches even the remotest areas of Africa). Jo, a tiny blonde not much bigger than the Bushwomen, is articulate and bright, and a natural mother. She loves children the way her in-laws love animals, and the Harnas preschool for her children and the Bushmen's children was her creation.

Marieta, Melanie, Jo, and Schalk settle around the dining room table, each with a scheduling book. As Harnas Wildlife Sanctuary becomes larger and more complex, these meetings have become more important. The four compare schedules for the day, making sure all the work will be done and yet nothing will be duplicated. They discuss updates on projects in progress, volunteer problems and situations that need to be addressed, general budget talk, and scheduling for the future. Frikkie is not in attendance because he is having the same kind of meeting with his volunteers over at the Volunteer Village.

The family's discussion is held mostly in Afrikaans, the first language of everyone in the family and many of the people living and working on Harnas. But they occasionally use other languages, especially English. Many English words have crept into the regular Afrikaans conversation: "printer," "email," "cocktail," "tourism," "budget" and of course the British expletive "bloody"—followed by whatever or whoever is driving someone crazy. In a passionate family as a whole, Marieta's language is the most colorful and full of emotion. Even if you don't know Afrikaans, you can sense her mood by the number of "fooking"-thises and "bloody"-thats she sprinkles into her regular comments.

Despite the business discussed in these meetings, Marieta is bottle-feeding and burping Ita, afterward letting her down on the floor, where she plays with the ears of a patient dachshund. Aviel crawls up onto Jo's lap, sleepy still, and occasionally demands her mother's attention. Samar breaks in with questions about his schoolwork, and the pleading eyes of various larger dogs fill the windows. "Why have you kept us out? You do this to us every day?" they seem to accuse. The Bushwomen move in and out, cleaning, arranging, fixing. It is hardly a typical CEO meeting, and yet on Harnas, it yields the same result—coordination, details, economy, and production.

Mara, the schoolteacher, sticks her head in: "Where are the keys to the school? They're not on the board! I need them to set up for class!" Another volley of Afrikaans between Marieta and Mara follows, and Mara departs, apparently having a better idea where the keys might be.

Having deposited Ita in a cozy enclosure with some older baby baboons, Marieta walks through the courtyard, trying to ignore Ita's "mrroooo, mrroooo"—the sound of sadness and abandonment that baboons use to call their mothers. Marieta heads for the food-preparation area, passing through gate after gate, each with its own particular form of lock or latch. Each enclosure on Harnas was added as needed, with whatever materials were at hand, so no two gates on the farm open the same way, but Marieta flicks them open and closed by memory. As she passes a large cheetah walking loose in the garden, she absentmindedly scratches his head, and he purrs in response. This is Goeters, the oldest cheetah on Harnas—23! Even though most cheetahs live only 3 to 6 years in the wild, they can live 20 years in captivity, and Goeters is the oldest known. He has been at Harnas since he was a year old, when he was found with his brother, staggering alongside a road. Today he wanders the garden much like a tame dog, and everyone who walks by him runs a hand along his back or scratches him under his chin. His purr is constant and forceful. Goeters appears in the pictures with Harnas's official patron, Angelina Jolie.

When he was young, Marieta tried to release him into enclosures with other cheetahs, but he didn't seem to realize he was one of them. He stopped eating and spent his days at the fence, calling to the van der Merwe family, the members of which he considered as his coalition—as a group of cheetahs is called. The vet advised the family to take back their newest "child," so he has lived free in the garden ever since, enjoying his status as the symbol of Harnas.

Cheetahs' faces can be distinguished from leopards' faces easily, as they have black "tear marks" running from the inner corners of their eyes to the corners of their mouths. The Bushmen tell the myth of a mother cheetah who lost three of her cubs at birth. The final and fourth cub lived, but one day it disappeared. The mother searched endlessly for her cub, crying and calling, calling and crying, her tears flowing so intensely that their trails stained her face dark in the line where they ran. Ever since then, all cheetahs have this tear mark in remembrance of that mother cheetah's great love and loss.

Cheetahs' bodies, too, are very different from other large cats. Built more like greyhounds than leopards or lions, they are made for speed, reaching up to 65 miles per hour—but only in short bursts of 20 seconds or so. If they can't catch their prey during that sprint, they have to stop and rest. Their claws don't retract, working like spikes on a runner's shoes, and their tails are long and function like a rudder on a boat, helping the cheetah to balance and turn at high speeds. After bringing down and strangling their prey—something they succeed at only 20 percent of the time—cheetahs usually have to rest before eating. During this critical rest time, cheetahs often lose their prey to other animals, like hyenas or leopards, because they are too tired to fight them off. They are beautiful but fragile creatures, and only about 10 percent of cheetah cubs survive to adulthood. Although they are a "protected" animal in Namibia, a loophole allows anyone to kill a cheetah if it poses a threat to people or livestock.

But Goeters doesn't know all of this. He just knows that he's happy, surrounded by those he believes are his own kind, his family. When visitors and volunteers first come to Harnas, Goeters is the first (and sometimes, in the case of visitors, the only) wild animal they get to touch. It is an amazing experience for people who have spent their whole lives afraid of even touching strange dogs. To hear, "*Go ahead and let him smell your hand. Then just reach back and pet his neck and shoulders,*" leads to an encounter that people will remember forever, especially once Goeters inevitably begins to purr or if he licks someone's hand with his sandpaper tongue.

Today, though, no visitors have arrived yet, and Goeters is pacing between the gate to the food-preparation area and the gate to the courtyard, waiting patiently for his bowl of meat to arrive. He watches other bowls of raw meat pass by, held high above the heads of volunteers carrying them, on their way to other cheetahs, leopards, lions, hyenas, and bat-eared foxes. He trusts that his is coming soon and doesn't even growl. Everyone who passes by greets Goeters and easily runs a hand down his spine because his back is at the perfect human-hand level. He won't stop long enough for more than a brief scratch until his food arrives. He casually rubs up against Marieta's leg while she surveys the activity around the garden, checking for problems or things out of place.

Marieta keeps a firm eye on the accounts of Harnas, so she occasionally makes inspections of the food-prep area. This morning the trailer with food for the animals in the outer enclosures—fully grown baboons, lions, cheetahs, leopards, vervet monkeys, caracals, and wild dogs, who live their lives, except in cases of feeding and illness, without human contact—is waiting for Etosha, the tour guide.

"Etosha! There is not enough chicken this morning. Give what we have to the volunteers for the cheetah cubs and replace yours here with some meat from the freezer room. And there should be more of this," she points to one large bucket holding red meat, "and that is too much for them," and points to another bucket. Every ounce of food—meat, fruit, vegetable, porridge—is accounted for, and every animal must get precisely the right food in the right amount.

When Marieta's husband, Nick, died in 2001, she knew almost nothing about the business end of Harnas, so after surviving that hard lesson, she keeps track of every penny spent on the farm.

She repeats her scrutiny with the food for the animals in the inner enclosures that the volunteers are preparing: "Too much here, not enough here, exchange this food for that." The volunteers scramble to make the changes.

During this inspection, Marieta somehow finds time to open the small shop she runs for the Bushmen's families, selling three loaves of bread to a man who was off the farm when the shop was open the day before. She never stops moving, never stops taking care of problems, always watching how her animals are treated because despite how large and complex the organization is now, Marieta still thinks of every animal as her baby. And every animal still looks to Marieta as "Mum." She can walk into enclosures of baboons she raised 15 years ago, and they gather tamely around her, drawn to her and hoping she'll take time to play. She can call each one by name.

Finally with a moment to herself, around 9:30, Marieta returns to her kitchen and sits down for some breakfast porridge, heated up in the microwave. Even then she is besieged with requests that, once she stops moving, catch up to her like a swarm of mosquitoes. Melanie asks for the keys to the office in the *lapa*— the outside eating area, lounge, and bar. Marieta swings the keys from around her neck and hands them to Melanie. Before long Marieta will be trying to remember where those keys have gone.

"When I die," she jokes, "I don't want a tombstone. I want a big key board, and everyone who comes to visit my grave will hang a key on one of those hooks instead of bringing flowers!"

Schalk appears, carrying his blowgun and darts, and Samar and Aviel scamper behind him. "I think we must dart Sarah and take the baby, Mother." She nods and follows him out of the kitchen, having eaten only a few spoonfuls of her porridge. She grabs a brightly colored scarf off a hook and wraps it around her waist like a belt. They walk through one gate, then another, and out onto the lawn, walking toward the teenage baboon enclosure. Along the way, she takes a second to nuzzle with Klippie— her baby giraffe—and Izzy—a baby springbok who comes only to Klippie's knees, both of them grazing—one on the leaves of the trees and one on the grass below. They make a strange duo, and yet they have become friends and roam around the lawn area together, often with their other companion, Jasper the donkey.

Jasper has been excluded today, though, forced outside the lawn gates because he has recently developed the tendency to bite people when he wants attention.

Halfway across the lawn, the office manager at the time, Juan, catches up to the group.

"I just got a phone call about some cheetah cubs," he tells Marieta. "A farmer found them on his property and he says he will shoot them if we don't come and get them."

"Have they been caught or are they still loose on the farm?" she asks.

"They are still loose, but he's trying to lure them into a cage."

"We cannot go out to every farm that has loose wild animals on it!" she reminds him. "It could take days to catch them. Tell him he must catch them first and then we'll take them." Juan turns and trots back to the office with his instructions.

Over at the baboon enclosure, several Bushmen are waiting at the gate to help with the darting and removal of Sarah and her baby.

"We separated Sarah from the rest of the baboons last night," Schalk tells his mother. "We put her in a smaller enclosure within the larger one when it became clear she was going to give birth soon."

Separating expectant mothers is done routinely to protect the mother and the baby from the rest of the troop—just in case play gets boisterous. Generally, baboons are a communal species with a complex social system, and child-rearing responsibilities are often shared, but exceptions to this rule have made the people at Harnas cautious, and every new baby is protected vigilantly. Schalk has already loaded a dart into his blowgun, and with Sarah in the smaller cage, he can easily make his mark.

He puts it to his mouth, and expertly, in one short burst, scores a direct barb to Sarah's hindquarters. Samar and Aviel look on with interest. Already by age five, Samar has observed more veterinary care than most adults do in a lifetime, and Aviel is also learning quickly. Aviel is still unsure about the blowgun, though, and she looks up nervously at Marieta.

"We're doing the right thing, Aviel," Marieta reassures her. "Sarah needs our help and we're not hurting her. We're only putting her to sleep for a while." Aviel nods but doesn't look convinced.

Once Sarah falls asleep, the Bushmen enter the enclosure and stand around Sarah's cage so that Schalk and one other man can put her on a stretcher with her newborn—premature, they can tell—and move her out of the large enclosure. Quickly, they carry the two baboons, still connected by the umbilical cord, across the lawn and through the gate to the animal clinic. Samar and Marieta, holding Aviel in her arms, follow, and when they arrive, they see that Schalk has prepared the room for the brief procedure of cutting the cord.

Two Bushmen lift the young mother baboon from the stretcher and place her on the table, and another lifts the tiny female, which is moving but weak. Blood is still flowing both ways through the cord, so Schalk takes two pieces of string and ties two knots—one about an inch from the baby and another about three inches farther along. Samar looks on with interest, but Aviel turns away.

"I'm doing this to help the baby," Schalk assures her once again, but it's clear that she is unconvinced.

"Put me down! *Ouma*! Put me down!"

Marieta plants a quick kiss on Aviel's forehead and stands her on the floor.

One quick cut in between the knots, a small spurt of blood, and the baby is free. Within a minute, the placenta is expelled from the sleeping Sarah, and the process is complete. Though the baby is small, Schalk feels confident that with Marieta's care, she will be fine.

Sarah was also once Marieta's baby, hand-raised in her home, so she first makes sure Sarah is all right before she takes the baby from Schalk and wipes her clean with a towel. Then she wraps her carefully in the scarf she has been wearing around her waist, creating a pouch for the baby. She hugs the infant to her and walks to her house and into her kitchen, leaving Schalk and the Bushmen to make sure Sarah is safe until she wakes up. Samar and Aviel skip off, looking for the next adventure on their way to school.

With the baby secure in the scarf, Marieta has both hands free to fix baby formula. She uses just the right combination of ingredients for a baby baboon. Any nipple is too big for this preemie, so Marieta uses an eyedropper to drip formula into the baby's mouth. She's in no hurry now—all the other chores can wait while Marieta fulfills the primary goal of her adult life: nurturing baby animals.

Once the baby baboon has been fed, Marieta looks down lovingly at the pink, almost-human face, now sucking on her human mother's smallest finger, having filled her tummy with all the formula she could hold. One of the baby's hands, with long slender fingers and tiny black nails, is wrapped around another of Marieta's fingers. "What a miracle. What a blessing!" she thinks. "Grace. I think I'll call you Grace." And Marieta leans her head back on the ravaged couch and smiles, her newest baby in her arms. It's just another morning at Harnas Wildlife Farm, where everything is unexpected and magical.

Baby Marieta with her mother, Anna

chapter two

LOVE *and* DEATH

Africa is a cruel country; it takes your heart and grinds it into powdered stone—and no one minds.

—Elspeth Huxley

MARIETA NEVER IMAGINED SHE WOULD BE THE CARE-taker for nearly four hundred wild animals in a sanctuary she herself would create and manage. She didn't foresee herself sleeping with baby animals, feeding them, taking care of their medical needs, and putting every penny she had into their survival. Instead, she assumed that she would carry on the work of generations of her family on their land in northeast Namibia, on the savannah at the edge of the Kalahari Desert. She would be a cattle farmer.

On the day her parents were married, her father, Schalk Willem du Plessis, forced her mother to make a difficult promise: they would wait 19 years before they had a baby. These years, he explained, would give them time to get to know each other and establish their farm without the financial burden and emotional

distraction of children. Schalk had lived for a time with his siblings, their spouses, and their children, and he knew how hard it
could be on a newly married couple to have children. Although
her mother longed for a baby, a husband's word was final in
that traditional and conservative Namibian culture, so Marieta's
mother, born Anna von Solms, abided by this condition.

After 19 years, as promised, Anna became pregnant and gave
birth to Marieta on June 26, 1950. To the sorrow of both parents, she would be their only baby, the only time Anna was able
to conceive. Since fathers with many jobs to do on a large farm
need help, Schalk desperately wanted a boy first and then several children more. But with no further children on the horizon,
Schalk made Marieta into the boy he needed on their remote
farm, called Tennessee. He raised her to be strong both physically
and emotionally.

As soon as she could walk, Marieta went everywhere with her
father, learning how to care for the cattle and other farm animals
like goats, chickens, pigs, and horses. She performed hard, physically challenging chores: mending fences, lifting bales of hay,
wrangling calves for branding, driving sheep, counting stock,
and helping with basic veterinary care. Her father encouraged
toughness, forbidding her to cry, saying she should be strong and
show that strength by keeping her emotions in check.

Anna, however, countered this hard labor and tough demeanor
by spoiling Marieta when she was in the house, even though she
was proud that her daughter was turning into such a strong, willing worker. During Marieta's childhood, her mother bathed and
dressed her and brushed her hair as she would do for a baby.
She had waited 19 years for this little girl, and knowing Marieta
would be her only child, she lavished all her attention and love in

an attempt to compensate for the hard life her husband pressed upon the child.

The family was shattered early one morning in May. In Namibia, this month marks the beginning of winter, when grass and seeds are at their most plentiful. Because Anna had asthma, the doctor had warned her that the family should move to Gobabis, the closest town, for the winter months to help her breathe more easily, but because none of the doctor's previous advice had worked for Anna, this recommendation was easy to ignore. First the doctor had said the trees on the farm were caus-ing her asthma, but when Schalk cut down the trees, she showed no improvement. Then the doctor blamed the horses, so Schalk sold their horses. Then he pinpointed the grass, but everyone knew nothing could be done about the grass. Finally, the doctor said Anna's asthma was caused by the many dogs that lived with the family, went everywhere with them, and even slept in their beds. Upset, Anna declared, "Then I will die because I can't give up my dogs."

"My mother's asthma was a constant in our house," recalls Marieta, "and I was always afraid of the next attack. When I came home from school, I would look for the family's car as I rounded the gate onto the property. If the car was there, everything was all right. If the car was gone, it meant Mother was in the hospital again. The car was gone more times than I could count."

On that May morning, Marieta woke up at four o'clock and heard deep, labored wheezing coming from her parents' bed-room. She got out of bed, checked to see that her girlfriend Babs, a houseguest from school, was still asleep, and then hurried into her parents' room, knowing she would find her mother strangled by yet another full-blown asthma attack.

"I found Mother lying on her bed, her lips bluish, her face pale and a look of pure terror in her eyes. She was pulling her hair out in clumps and tearing at her nightgown around her neck, grasping for the air she couldn't reach. Her fingernails were making red streaks down her throat.

"My father handed me the pump filled with adrenaline and told me to get busy. I pumped the medicine into my mother's mouth, and the shushing sound was added to my mother's gasping and choking. I pumped and I pumped until my father took the pump from my hand and shouted, 'Go and get the injection!' I ran to the cupboard but didn't know exactly where the syringe was, so I pulled out all the drawers at once to find the needle. I had been asked to be a nurse many times, so I knew how to fill the syringe with the right dose from the little bottle of epinephrine, a stronger medicine than we had been pumping into her mouth."

She hurried back to the bedside and watched while her father jabbed the needle into her mother's hip. Then they waited, hoping for improvement. Gradually her mother's breathing eased. Both father and daughter let out long sighs, and Marieta took her mother's hand in one of hers, while the other smoothed the damp hair back from her mother's forehead. In silence she stroked her mother's slack arm and tried to soothe her tense forehead and tightly closed eyes. Exhausted, Marieta let her head fall forward on the bed next to her mother and dozed off.

She was awakened some time later by the same sickening sound as before—her mother struggling to breathe. Marieta shook her father, who had fallen asleep in a bedside chair, and their panic soon returned.

"Go get some hot water," he ordered Marieta. "We'll put some Vicks in it and make her breathe it."

They didn't have an indoor faucet with hot and cold running water. Marieta had to go outside to the pump. She struggled with the stubborn handle, yanking it up and down, using her whole body as leverage. Once she had pumped enough water into the pot, she returned to the house and began to heat the water on the stove. Her mother's wracking, choking gasps continued, often escalating to a frightened scream.

"Hurry! Hurry! Hurry!" Marieta whispered, willing the water to get hot faster, every nerve in her body keenly aware of her mother's suffering. She listened as the screams settled into a deep rattle in counterpoint with her mother's labored breathing. Then suddenly—nothing. The sound cut off like a plug being pulled. A few seconds later, she heard pounding against the wall and frantic cries—this time in her father's voice.

Marieta knew what had happened. She turned off the stove and ran to the bedroom. She found her father banging his fists and forehead against the wall, raging and crying uncontrollably. Her mother lay on the bed, still, but with her eyes open and her mouth agape, as if she were struggling for one last breath.

"Fix her!" he shouted and turned away.

Marieta sat next to her mother on the bed, anxious about touching the still body but afraid not to do what her father had ordered. She gently closed her mother's eyes and lifted her jaw until her mouth closed. By instinct, Marieta slid her mother's wedding ring off her finger and onto her own. She didn't cry— she just sat, stroking her mother's face.

"I knew that even though Father was sobbing, I had to maintain control. And I had been taught to, after all. My father didn't know how to handle emotion in others. He would just shut down, and I needed as much of him as he could muster. I knew that especially now, I had to be the strong one," she recalls.

"Call the family," her father choked out, and Marieta headed to the kitchen where the phone hung on the wall. First she called the closest neighbors and asked them to come and help since they lived only 20 minutes away. Next she called her uncle Frans, who lived ten miles away. He and Aunt Martha were brother and sister to her parents, making their sons and Marieta double cousins, but as an only child, Marieta felt so close to the boys that she considered them her brothers. The two families had even lived together until her aunt and uncle bought their own farm. "I'll come immediately," Frans said.

By the time the neighbors arrived, Marieta's friend Babs was awake, having slept through the entire crisis, but she grew more upset by the minute as she learned what had happened, so the neighbors offered to take her home. Exhausted and dazed, young Marieta sat on the couch while neighbors and family fluttered around her.

Later that day, Marieta accompanied her uncle to Gobabis to buy a coffin because her father was too distraught to think clearly. No one in the family talked much, so the hour-long journey was a quiet one, giving Marieta time to remember the difficult night and to consider what lay ahead of her now that her mother was gone. Picking out the plain, unadorned coffin was an easy task, as in the small town of around 4,000 people, choices were limited. Uncle Frans and the undertaker slid the coffin into the back of the truck and they drove it back to the farm.

By the time they returned home, the neighbors had all left, and because Aunt Martha was home taking care of her four boys, ages 12 to 16, this left Marieta the only female in the house. So she was given the chore of washing and dressing her mother's body. Her father and uncle remained outside the bedroom, leaving this "woman's work" to young Marieta. She struggled with her mother's body, heavy and cumbersome in death, and the girl was, of course, in shock. She undressed her mother, carefully removing her nightgown, torn in her struggles. The child used a soft cloth and warm water, gently bathing her mother's body that had suffered so much in her last night on earth.

Marieta had had little exposure to human death at this point in her life, and all she could compare it to was sleep. Accordingly, she dressed her mother in her nicest nightgown, struggling to turn, lift, and straighten it. Somehow she completed the washing and dressing, and then called her uncle and father. The two men carried the body to the waiting coffin, nailed down the top, and loaded it back into the truck. The two of them left Marieta at home while they drove to Gobabis where the doctor would certify the death and her mother would be buried. And with that, Marieta became the woman of the household.

She was a month shy of turning 12 years old.

She told herself that she wouldn't cry—in fact, she promised herself that she would never cry again. She knew she had to be tough—for herself and for her father. Life continued on the farm, with Marieta's responsibilities increasing both inside and outside the farmhouse. "I got good at working with animals," she says, "and while I learned about them, I learned to love them." At this point her "pets" included only farm animals and the numerous dogs and cats that the family kept. "My first dog, Lissie, spent every minute with me, both day and night. I played with dogs the way other children play with each other. Lissie and I had tea parties, acted out pretend-school, and played dress up."

Animals were always drawn to Marieta and would willingly submit to being playmates. Marieta admits, "Even now I'm more of an animal person than a people person. I feel more comfortable around creatures. With people, I prefer to stay in the background, letting others do the talking. That's one of the reasons going to school was harder for me than for most."

Direct contact with wild animals was limited to a single encounter when she was only six years old. Her school still had an outhouse, and Marieta left class one day to use the outdoor restroom. When she opened the heavy wooden door, she found herself face to face with a huge, fully grown male baboon. An encounter with this aggressive creature was too much for the little girl, who was so scared she lost control of her bowels. Fortunately, this chance meeting did little to alter Marieta's love for animals of every kind.

People caused Marieta more grief than animals generally. In one tragic situation a neighbor put out poisoned meat—probably to kill local jackals that often stole livestock—and all of Marieta's dogs, including Lissie, found the meat, ate it, and died horribly painful deaths. This happened the same year that her mother died, and Marieta was inconsolable. To help ease the pain of these losses, some family friends from South Africa gave her a vervet monkey as a pet. She named him Oscar.

The choice of house pet was not as exotic as it would seem. Living in Namibia on a remote farm, Marieta regularly had the opportunity to observe wild animals like vervet monkeys, wandering cheetahs, baboons, springbok, duikers, and oryx. Even though the vervet monkey was the first wild animal Marieta had owned, the idea of having one of these animals as a pet seemed quite natural to her.

In those days, Namibia still had wide-open spaces without fences, and nomadic peoples still hunted the free-roaming wild creatures for sustenance. In her spare time, Marieta wandered the bush, too, and she grew to appreciate the wonder of wildlife. Some of the wandering bush people—a tribe called Nama—worked for her father, and Marieta became fairly proficient in understanding their musical language with its clicks, clacks, and ticks, learning to speak some as well. Their children became her playmates, and the savannah their playground. Seeing wild animals while the children were playing was a normal part of growing up on a farm.

The same independence she experienced in the wild affected how she interacted in social situations. Rather than run to her parents with her problems, Marieta learned to settle them herself. Her father encouraged this toughness. If he found out, for example, that she was having a problem with another child, he would set up a fistfight for her, clearing furniture to one side and ordering the two to "Hit each other! Fight!" Marieta became an accomplished fighter, beating boys larger than herself, and her reputation for fierceness was well known among her friends and classmates at school.

She needed it during the week. Unlike in America, it is common for white children living on farms in Namibia to attend boarding school. As early as age six, farm children whose parents can afford it are sent off to "hostel" from Monday to Friday, usually returning home to their families for the weekends. After her mother died, Marieta returned home only every third weekend. This separation may seem harsh to American parents, and to be fair, Namibian parents don't like it either, but a good education requires it, and most children know nothing else. Attending boarding school is just the norm for farm children.

When Marieta first went off to hostel, Anna asked an older family friend who was a student there to watch out for Marieta. When they dropped her off the first time, Marieta cried and cried, but the older girl said, "Come to me! I will take care of you!" She cried some more, but moved into the arms of the new friend. Once little Marieta's mother drove away, though, the girl slapped her and yelled "Shut up! No more crying!" The situation only grew worse. Marieta remembers the girl lying on her bed, ordering Marieta around. Once she said, "I want water! Bring it to me by the time I count three!" So Marieta ran to the kitchen, but while she was waiting for the water to fill the glass, she heard the older girl yell "Three!" When Marieta returned with the water, she was rewarded by being beaten with a coat hanger.

Marieta would suffer indignities at the hands of other girls in hostel as well as schoolworkers and teachers. Whippings and other physical punishment for real and imagined sins were often meted out. "One of the cooks had a whip with a wooden handle and a leather strip hanging down from it with pins at the ends, a weapon called a sheer strop," Marieta remembers. "When she hit girls with it, it left small red pinpricks, and I remember one time getting hit with it for cleaning out a pan that still had food in it. It wasn't my fault. Another girl tricked me into throwing out the food, but I was the one who received the whipping."

Away from home, children learned to be independent and self-sufficient, and as she got older, her upbringing as a tough tomboy came in handy. "I was a cheeky child," Marieta confesses, "who sometimes got in trouble by opening my mouth when I shouldn't have. I've always had a strong sense of justice, though, and more often than not, I was defending myself or an innocent classmate when I challenged the actions of a teacher or other adult.

"In high school, I had a German teacher who hit students freely for what he considered 'sins.' If a student wrote the wrong thing on her paper, for example, he would hit her very hard between the shoulder blades—often hard enough to make her fall over! He hit me a thousand times, but I was tough enough to take it because of the training my father had put me through. When the teacher hit one of my friends who was completely innocent, though, hard enough to knock her over, he pushed me past my breaking point. I picked up my books and schoolbag and walked to the door."

"Where are you going, young lady?" he bellowed from across the room, where her friend was doubled over, her breath knocked out of her from the instructor's blow.

"I will not stay in this class anymore! You are not allowed to hit us like this!" And Marieta stomped out of the room.

From that point on, the teacher held a grudge against Marieta for blatantly disrespecting his authority, and he constantly looked for errors to give him cause to punish her. Other teachers followed suit, as word of a disrespectful child gets around quickly in a small school. One day in math class, a girlfriend of Marieta's copied her homework because she hadn't done it herself. Marieta had committed an error that showed up on both papers, so the teacher called them both to his desk.

"Off to the principal for you two cheaters," he said. When Marieta protested that she hadn't done anything wrong and so refused to go, he picked her up by the back of her shirt with one hand and the waistband of her skirt with the other and tossed her out the open window! She landed with a thud, unhurt but mortified at this disgrace. "Do not come back in here today!"

he shouted at her. "Spend the rest of the period running around the building!"

Off she went, running the length of the building and turning at the corner. But Marieta had a boyfriend in a classroom she would have to pass, and, eager to keep her humiliation to herself, she slowed down when she got to his classroom window and walked by nonchalantly, as if she was out for a casual stroll. As soon as she passed his window, she started to run again, repeating the slowdown each time she passed his room. She refused to show weakness to either her boyfriend or her teacher and ran for the rest of the period.

Because Marieta had been taught to settle her own disputes, she didn't report to her father either teacher's abuse, or the abuses by the older girl and the cook. But soon after the window episode, Marieta had her tonsils removed, and while she was waking up from the anesthesia, she talked nonstop, telling her father about all of her fights and problems at school. He knew she wasn't lying, and when she finally became lucid, he asked her if it was true. She nodded.

"Why didn't you tell me these things?"

"You told me to fight my own battles," she answered. "And you said I shouldn't tell on others."

Incensed by the abuse of his daughter and other students, her father went to her school even before Marieta was released from the hospital. He grabbed her math teacher around his neck and demanded, "What did you do to my daughter?" When the teacher admitted he had thrown Marieta out the window, her father punched him in the face and stormed out.

When Marieta returned to school, the teacher refused to have anything to do with her. He told her to sit in the back of the classroom, facing away from him. "You are no longer a part of this class. Do whatever you want but you are invisible to me. I don't care." While these experiences seem harsh, they only made Marieta tougher, a trait that helped her survive all that she would face in her later life.

"After the death of my mother," she recalls, "life with my father on Tennessee became unbearable. He just couldn't get through his grief, and taking care of a young girl was the last thing he could think about. I spent all week in hostel, but when I came home, he didn't know what to do with me—even how to feed me. One of his staples was pulling bread right out of the freezer, putting meat on it, and giving me a bowl of sour milk. We were not poor, but my overly frugal father refused to buy food." Shaking her head slowly, she concludes, "He was useless to me."

Eventually, his grief became more than he could bear, and he tried to kill himself with a shotgun but didn't succeed. Realizing that her father was beyond caring for her, Marieta moved in with her cousins for most weekends and some holidays. The four boys continued her education in how to be tough. One time her cousin Frikkie put her on a donkey, took a thorn from a bush and put it under the donkey's tail, and finally whopped him on his flanks. Off he went, bucking and kicking while Marieta held on for as long as she could. After being thrown off, she begged to have another go, sure she could stay on longer. The boys learned early that Marieta was different from most girls. She worked for and earned their respect.

Being a tomboy came easily for her, but becoming a woman was more of a mystery. At 14, Marieta started her period, but she had no idea what was happening to her. Her mother hadn't had a chance to explain menstruation before she died. In the mid-1960s, girls still didn't talk about private issues among themselves, and Aunt Martha—raising four boys and therefore never having to explain menstruation—probably assumed Marieta's mother had told her earlier. Marieta was left not having anyone she felt comfortable with telling about the strange changes that were going on in her body. Her first period happened during a school holiday while Marieta was staying with her father, and understandably, she was terrified. She thought she was very sick—even dying. The only remedy she could think of was to stuff toilet paper in her panties to soak up the blood. The blood kept coming, though, and Marieta was afraid to move around, afraid the blood would drip down her legs and be seen by her father.

She was so scared that she went against her father's mandate and broke down crying. Her father was baffled by her outburst. When she was asked to go out to a party that weekend, she refused and cried some more. All she did during that holiday was sleep, sit, and cry, afraid to do anything that might encourage the blood to flow. Her father was puzzled but reasoned she must finally be mourning her mother's death.

"What's going on with you?" he demanded. "Are you crying for your mother? Are you going crazy?"

Finally she admitted she thought she was dying, telling him vaguely that blood was "coming from underneath her."

"Oh my God!" he said, grabbing his head and shaking it in disbelief and shame that his daughter had been so neglected by them all. He called her aunt and told her to come immediately. Soon Martha was sitting with Marieta, explaining the changes her body was going through. Marieta was relieved—also surprised and amazed—that what was happening was natural, that she wasn't, in fact, dying. Her aunt gave her some pads, but in an oversight, forgot to tell her what to do with the used pads, since on a farm in Namibia, there was no regular garbage pickup. As the months went by, Marieta collected the used pads in a box behind her bedroom door. Ashamed of them, she didn't dare ask anyone what to do.

One weekend while she was visiting home, her father bellowed at her from her room. "Come here! What is going on behind your door?" he said, pointing at the box of used menstrual pads.

"It's the things I must use every month," she answered, unbearably humiliated by his discovery.

"You cannot keep these here! What are you thinking? You must take them out to the dung heap and burn them, every one!"

She was embarrassed but also relieved that she finally knew what to do.

Nothing she had experienced, though, prepared her for her father's remarriage when she was 16, to a woman named Gerty, who had five children. Marieta was at first elated because she finally had siblings—something she had always wanted. Two of the children were older and already out of school and on their own, but three were still in school, like Marieta. Yet these children were not like Marieta at all, in that they had not been raised to be tough. In fact, they were rather spoiled. They were completely different from her cousins, as well. In describing them, Marieta flutters her hands, saying "hoo hoo hoo," with her nose in the air. They didn't ride horses or play with animals or do anything physical. They were snooty and superior toward their new sister. She was clearly the outsider, and they never let her forget it.

"All they did was sit and smoke!" Marieta reminisces. "So of course we fought about everything—especially with the youngest boy about my age, Chris. When we'd disagree, my father made us fistfight in the kitchen. We'd fight, but I'd always win because that boy was so soft! My stepmother, always more protective of her own children than of me, would scream, 'I will call the police! Stop it! I will call the police!' But mainly she was upset, I think, because her son never could beat me."

Just like stepmothers in fairy tales, Marieta's new mother favored her own children over Marieta, and her stepsiblings tormented her whenever they had a chance. But in her typical way, Marieta refused to tell her father what was going on. When all the children went away to school, Marieta's father would give each of them a dollar, and then her stepmother would put extra money in the socks of her own children. She also sent candy and other sweets with them, but gave none to Marieta.

When something went wrong on the farm, Marieta's father would say, "Who did this?" The siblings—and her stepmother—would gang up on Marieta and say she was the guilty party. Marieta withstood this abuse stoically, refusing to tattle on them to her father, but she claims she hated her new family, especially her new mother.

"As an adult, my feelings changed, though, and when my father died of bone cancer, Gerty came to live with me. I took good care of her, and we finally became true friends. I even came to love Gerty," Marieta asserts. "When Gerty became ill, her own children refused to take her in, so she continued to stay here with me. When she started showing signs of Alzheimer's, they pressured me to put her in a home in Gobabis. I fought them on this because I knew I could take better care of her on the farm, and I wouldn't be able to go to Gobabis very often to visit her. Her own children wouldn't visit her, so in Gobabis, she'd be completely alone."

But Marieta was not her biological daughter, so her children won that battle. They moved their mother to a home where she was completely alone and at the mercy of the workers there. Marieta tried to visit as often as she could, even after her stepmother stopped recognizing her. Gerty died soon after, as predicted, all alone.

Despite her early dislike of her siblings, they, too, became friends as they all grew up, got married, and matured into adults who moved past the petty jealousies and fights of childhood. Marieta says, "We are all good friends now. The youngest brother that I fought in the kitchen is now my good friend—married with three children and living in Windhoek. It's important to hold on to family wherever and whenever you can. Life is hard and messy, and toughness and forgiveness must go hand in hand."

Marieta grew into a tall, beautiful young woman with long, wavy blond hair, brilliant blue eyes, and a dazzling smile that charmed others, even though she claims she's not comfortable in social situations. Her confidence and sense of humor made her a favorite of many, and she never lacked for friends who appreciated her laughter and her joy in ordinary things. Her love for the land and its animals spilled over into love for others. People were drawn to her because she was a happy, genuine, straight-talking girl.

When boarding school was over at age 18, Marieta joined the police force in Gobabis, an unusual career choice for a young woman, but one that Marieta was suited for, with her confidence, her intelligence, and her fearlessness. She spent a year in the training school and two years on the force as an investigator, doing everything from filing to investigating murders. On the force in Gobabis, she ran into former classmate Nick van der Merwe, who was also working for the police force—although his assignment was in the north, in Ovamboland, where he was a member of an antiterrorist unit.

The Ovambo tribe is part of what is considered the "ethnic" population. This tribe, which has lived in the northern part of the land since the 14th century, makes up about 50 percent of Namibia's current population. In the late 1960s, their land along the Angola border became the base of operations for the South West Africa People's Organization (SWAPO), a group attempting to liberate Namibia from South Africa's rule through armed struggle. Independence finally came in 1990, and SWAPO became the ruling party in Namibia, responsible for the democratic, stable, and economically developing country it is today. When Nick was on the police force, however, SWAPO was considered a terrorist group, not the freedom fighters they are called in retrospect.

Marieta and Nick were well acquainted because they had dated each other's best friends in high school, although Nick was two years older. Marieta was the roommate of Nick's girlfriend in hostel, and many times she was called upon to pass love letters between them, and they often went out as a foursome. The two couples had broken up eventually, and since Nick had graduated two years before Marieta, they had lost touch. Marieta was dating someone else when she became reacquainted with Nick, but she remembers that boyfriend as being chauvinistic and demanding:

"He would yell that he wanted coffee," she recalls, "and I would be expected to jump up and get it for him. He wasn't a gentleman at all. I remember one time when we were going to a dance, and when I came to the door, he said, 'I don't like that dress. Go change into something else.'" Nick was a pleasant change. He opened doors for Marieta, treated her with respect, and told her over and over that she looked beautiful—no matter what she was wearing. She sums up her feelings about him with "I felt like a queen while I was with him." The contrast between the two young men was easy to draw, and so Marieta dropped her other boyfriend and started dating Nick.

Nick was a handsome, rugged-looking young man, and like Marieta, the only child of cattle farmers. His family owned many farms, and he had grown up much as she did, working hard, learning about animals—both domesticated and wild—and feeling the pressure of being his parents' only heir. Nick knew that one day he would be responsible for what his parents had built. For those two reasons—a cattle-farming background and the pressures of inheritance—they could understand each other better than most.

But Nick was more than a gentleman and a tough policeman. Marieta remembers him as a wonderful father who was gentle and kind with his children. "The only time he hit his boys was when he caught them smoking. I was known as the funny woman with the contagious laugh, but Nick had a dry sense of humor that caught people off guard. He was so quick! When you would least expect it," she chuckles, "Nick would say something, there would be this pause as everyone realized what he'd said, and then everyone would laugh."

Nick's job was much more dangerous than Marieta's because he was engaged in the struggle for independence raging near the Angola border. As part of an antiterrorist squad, Nick wore civilian clothes and worked undercover, trying to locate members of SWAPO and figure out their plans for attacks, bombings, and espionage. One night Nick and his force were attending a local dance—off duty, as civilians—when they got a call that there were terrorists holed up in a house in a nearby village. The team immediately left the dance and headed out.

"Nick told me there was no moon and it was extremely dark," Marieta says. "The unit parked down the road from the house and moved quietly through the bush, surrounding the house without ever being seen. Then they announced their presence with megaphones, saying, 'Come out! We have the house surrounded!' As they'd planned, the members of the unit all turned on their flashlights at once, aiming at the doors and windows.

"At this point things happened fast. One of the terrorists bolted from the house, running straight at Nick, whose large light caught the man directly in the face. Nick saw that the man was holding a gun and aiming it straight at him—Nick was an easy mark because his flashlight gave the man a target. Instinctively he turned his flashlight to the side and confused the terrorist, who turned his gun in the direction of the light rather than at Nick. This gave the rest of the men time to shoot the terrorist and bring him down. The man fired off some random rounds, though, and one of these hit one of Nick's friends on the force, Manie, paralyzing him for life."

Though tragedy could be expected for men fighting in a guerrilla war, this event affected Nick deeply. His father had never liked the fact that Nick worked in a dangerous job, especially because Nick was his only son. He clashed with Nick over quitting the antiterrorist force, even resorting to bribery. He told Nick, "If you come back to Gobabis and work on the police force there, I'll buy you a Mercedes." After some deliberation, Nick agreed, partly because he was still in shock over Manie's paralysis. So Nick began working in Gobabis, and there he met Marieta again. Before long their friendship turned romantic, and they were married July 24, 1970, when Marieta was 20 and Nick was 22. A year later, Marieta was pregnant with their first child, a boy they named Nicolaas Johannes Rudolph—calling him Nico.

Because money was tight, both Marieta and Nick continued to work on the police force. Nick's cases in the north, which he had to close out before he could be free of his former position, continued, forcing him to make numerous trips away from home. Then the birth of Nico led to an unexpected complication. During and after Nico's delivery, Marieta bled a great deal, and the doctor didn't diagnose this anomaly. He simply sent her home. During her first week of motherhood, Marieta struggled with pain and occasional bleeding. Nick had to leave for work in Ovamboland soon after Nico's birth, so Marieta stayed with her father and Gerty so they could help with the baby.

Ten days after Nico's birth, Marieta began bleeding heavily. Gerty tried everything she knew, but she couldn't stop it. Blood and more blood, and soon Marieta went into shock from the loss. Schalk and Gerty drove her quickly to the small hospital in Gobabis where the doctor diagnosed her with septicemia—blood poisoning.

Marieta had lost so much blood that the doctors believed she would die. They sent a helicopter to retrieve Nick from the border so he could be with her one final time. Meanwhile, one of the doctors treating Marieta—who had the same blood type as she did—gave his own blood to keep her alive. He lay alongside her and transferred his blood directly to her. They sent for more blood of Marieta's type, but the doctor's sacrifice probably saved her life. Eventually, she began to show signs of recovery.

Marieta survived, but she was advised to stay in the hospital for an entire month. Nick returned to work, so Nico lived with his grandparents while Marieta recovered. She returned home at the end of that month, but the bleeding started again, so she went back to the hospital. When she finally recovered, she went to her parents' house again to retrieve Nico.

"They handed me a baby I didn't even recognize because he had grown so much since I'd seen him. I pulled back the blanket and saw Nico looking up and I was amazed! I had given away a newborn and I got back a big baby who was already smiling," she remembers. "It was wonderful and at the same time it broke my heart to think of how much I had missed in the first months of Nico's life."

While Nico was an infant, Marieta could take him to her office and he would sleep through most of the day, but when he started walking and then running, she could no longer have him with her. First, Marieta hired a friend to take care of him, but Nico hated having his mother leave him every morning. As soon as she started to dress him, he began to cry.

"Eventually, my girlfriend couldn't work as his nanny, so I hired a Bushwoman," Marieta recalls. "Nico cried even more when I left him with her, and soon I found out why. The woman would hit Nico when he didn't do what she wanted; then she would hit him when he did do what she wanted, and she would hit him for no reason at all! Once I realized what was going on, I fired the Bush-woman and got a full-time nanny. Nico didn't seem any happier, though, and the guilt I felt over leaving him made me miserable."

Finally, Marieta fired the nanny and made a deal with Nick's parents to take Nico full time. It was too far to drive out to their farm every morning, so Nico had to stay with them for the whole workweek. Every Friday afternoon, Nick and Marieta would drive to the farm to spend the weekend with him. The stress of leaving her toddler son over and over, though, became too much for Marieta. The couple knew it wasn't fair to Nico for his mother to be away all week and his father to work far away on the border, coming home for only a few days at a time.

Both Marieta's and Nick's parents were called farmers. In Namibia, however, this usually means something different than in America, where "farming" means raising crops like corn or wheat. In Namibia, farming generally means the raising of animals—cattle, mainly. (They don't use the word "ranch," which is a New World term borrowed from the Spanish/Mexican "rancho.") In Namibia, wealth is also counted differently. Wealth isn't having money in the bank but rather cattle on the land. The more head of cattle, the more wealth. Nick's parents were extremely wealthy by this standard, owning many cattle farms, and Marieta's parents were nearly as rich.

Knowing they would both eventually inherit their parents' farms, Marieta and Nick decided to give up working on the police force and start farming immediately. The couple's parents did not believe in spoiling their children. Giving them a farm to begin their working lives—even a farm they would eventually inherit— wouldn't teach them anything about self-sufficiency and responsibility, and the young couple never expected such a gift. Instead, they worked out a deal to buy a farm from Nick's father, and they went to a land bank and took out a loan to pay for it.

The farm they bought covered about 8,000 hectares or approximately 20,000 acres. Located northeast from Gobabis, near the Botswana border, it was called Harnas.

Elsa relaxing on the family's sofa

chapter three

BABIES

Never doubt that a small group of thoughtful, committed individuals can change the world. Indeed it is the only thing that ever has.

—Margaret Mead

"STOP, NICKY! STOP!" MARIETA EXCLAIMED. "LOOK AT that boy! He's selling some kind of animal."

Nick obligingly slowed the car and made a U-turn. They were driving from Harnas to Ovamboland, where Nick still had one more case dragging on—even though he had quit the police force several years earlier. Marieta was going with him this time to see friends in the region, and they were passing through a small village on the way.

A child selling something along the side of the road is a common sight in Namibia. The roads near villages are often lined with children trying to sell things to white people passing through. This boy was holding a baby vervet monkey, the same breed as Marieta's first monkey pet, Oscar. The animal was undersized and miserable-looking. Vervet babies are extremely dependent on their mothers for months after birth, clinging tenaciously to their chests, so Marieta realized that this monkey had either been orphaned or taken by force from its mother. Either way, she couldn't stop herself from rescuing it from the careless clutches of the young boy.

"I want this monkey," she told the boy in Afrikaans as she climbed out of the car, but he spoke only Ovambo and enough English to say, "I want money."

Marieta looked at him, noting his poor clothing but knowing that giving money to a child often meant it would go to his father, who would likely spend it on alcohol.

"I have a fresh loaf of bread," she said and took it out of the backseat and extended it to him.

He eyed the bread, then refused. "I want money."

Marieta again considered the poor, frightened monkey clinging to this small boy who wouldn't know the first thing about taking care of it. No doubt the monkey would die in a few days from malnutrition, neglect, or worse. She pulled out five South African rand—the equivalent of less than a U.S. dollar—from her bag and handed it to him.

He nodded and handed over the squealing infant to Marieta. It was a female with oversized sad eyes, red and filmy from malnutrition.

Marieta and Nick had no idea how this one exchange for a monkey would alter their lives forever. This transaction in 1978 signaled the beginning of Harnas Wildlife Foundation and the end of cattle farming for the family, although they wouldn't see the completion of that change for over 20 years. And Nick would live to see only the beginnings of the magnificent sanctuary that Marieta and her children would build.

Marieta named the monkey Adri—one in a line of troubled monkeys to be rescued by Marieta that she would give that same name—and took her home. By now the couple had three young children, so Marieta had plenty of bottles to feed Adri milk, but she somehow knew that the monkey needed something more substantial. She was too tiny and clearly undernourished.

Some things can't be learned from a book or even from attending veterinary school when dealing with injured and abandoned animals. Instead, Marieta relied on her instinct. She fixed a pot of porridge with some baby milk and put a bit in a teaspoon. Adri grabbed the spoon and ate the porridge ravenously. Marieta repeated this action over and over until the monkey was full. Then Adri fell asleep, tightly clinging to her new mother. That night Adri slept with the van der Merwes, the first night of many with monkeys—and other wild animals—in their bed, right alongside the dogs and the children. Fortunately, the bed was a large one.

The family had been living at Harnas for four years. At first they had no sources of income, no cows, no sheep, nothing but a piece of land with some water—and a mortgage to pay each month. Both Nick and Marieta received small pensions from their work on the police force, and with this stipend they started to put a roof over their heads. Marieta's father finally showed pity on the struggling couple and gave them a few cattle and sheep, and Nick's mother offered some trees from her land, which they hired Bushmen to dig up and replant on Harnas.

Nick and Marieta built their own home, brick by brick. Marieta remembers those difficult times of hard labor: "I made the bricks, adding together the right amounts of dirt, straw, and water, and mixing it to a biscuit-dough consistency. I'd shovel the concoction into molds and lay them out in the sun to dry. Then I'd take the ones that had spent the past few days drying into hard bricks and carry them to Nick, who put them in place with mortar, slowly building our one-room house." Even when she was pregnant with their second son, Schalk, Marieta made and carried the hardened bricks to Nick: "I still thought of myself as a newlywed and tried to look good for my husband. I was hugely pregnant, and carrying bricks—all the while wearing a miniskirt!"

They built one room and a bathroom, and as they'd squirrel away a little more money, they'd build another room onto the house. In this way, they managed to create a home—although they continued to drive to Nick's parents' house for dinner because they didn't have enough food to feed their family.

The family grew quickly. Schalk was born in August 1976 when Nico was four. A few months later, Marieta felt exhausted and nauseated, and so went to the doctor. He told her she was pregnant again, but she argued she couldn't be because she was taking birth-control pills.

"Well, there was only one Jesus born," the doctor replied, "so you must have done something wrong. You're definitely pregnant."

Marieta went home and thought hard about what could have happened. Finally she talked to Kasoepie and learned that as Marieta used the birth-control pills Kasoepie had been replacing them with baby aspirin—which looked virtually the same. Kasoepie thought the pills were vitamins for the two boys and believed she was doing Marieta a favor by keeping the packet full. Marieta had been so preoccupied with her husband and two sons, building a house, and taking care of her animals that she hadn't noticed that her prescription never seemed to run out.

Eleven months after Schalk was born, she gave birth to their daughter, Marlice. She was born prematurely and was so small that the doctor nicknamed her *vlooi*, Afrikaans for *flea*. The nickname stuck, and today many people don't even know Marlice's real name. Marlice grins, "Even newspaper articles about me call me Vlooi, and when I call people on the telephone, I have to introduce myself that way or no one will know who I am."

Two babies on bottles and two babies in diapers: perhaps this was nature's way of preparing Marieta for all the baby animals she would care for later in life.

The family sleeping together in the same bed while the children were small is a normal arrangement for people in Namibia and in many other countries as well. Marieta can't imagine it any other way and remembers the cold winter nights when those "sweet little children" of hers would "snuggle up and everyone would keep warm together." Even before the children, their dogs always slept with the couple. When they were first married, Nick had only one animal, a German shepherd, who was protective of his master and jealous of Marieta.

"As I got ready to get into bed," Marieta recollects, "I would shout 'Nick! Here I come!' And he would grab the dog and hold his muzzle. The dog would growl and fight until I jumped into the bed, and then he'd settle down." Marieta had several dogs of her own, which also slept with them. Then along came the children, and soon after, the monkeys, baboons, and other creatures. Her bed was never a lonely place.

At first when they went anywhere, they loaded into the Mercedes every person and every creature that couldn't be left alone. Then they would drive to their parents' houses, to friends' houses, or to Gobabis for shopping, carrying babies and baboons, walking along the streets and into the stores, and leading numerous dogs on leashes. "Eventually," she laughs, "we had so many animals and children to transport around that I gave Nick an ultimatum: 'Sell your bloody Mercedes!'" Nick, peacemaker and pleaser of the wife he loved, complied, and the family bought its first "bucky"—or truck—to carry all of Marieta's loved ones.

When the family went shopping in Gobabis for supplies, Adri would perch itself on Marieta's shoulder, so of course people would stop and ask them questions. Word started getting around that Marieta van der Merwe would take in orphaned and injured animals. Because so many animals were shot to protect livestock—often leaving baby animals orphaned—there were a lot of animals left wounded on the roads, and so many animals were trapped and removed from habitats now controlled by people, that people were relieved to find someone willing to take these animals in. Without this option, farmers usually had no choice but to shoot them.

Marieta's second rescued animal came as a result of a call from a couple who had seen her with her animals in Gobabis. They had a baboon who was tame and fully grown. They couldn't keep her anymore, they said, and wondered if Marieta would take her. The baboon had spent her whole life living on a chain, and at first Marieta kept her the same way—although a very long chain—simply because the baboon didn't know Marieta, and at this point Marieta didn't know much about enclosures and environments for animals. She soon figured out, though, that the baboon needed her own enclosure, where she could climb trees, play, and be free to roam in a more natural setting.

Throughout the years, Marieta has learned more and more about what specific animals need in each enclosure, how big that enclosure should be, and how many animals of each species can live together peacefully. Leopards, for example, are solitary creatures except when they come together for breeding, so each needs a separate enclosure. Baboons, on the other hand, live in troops ranging from just a few animals to three dozen. Lion enclosures usually include one adult male and one or two females.

Calling an enclosure a "cage" doesn't even come close to describing where the animals live on Harnas, although usually chain-link fencing is used around each perimeter and sometimes electric wires keep some animals—like baboons—from climbing over the fence. Unlike a zoo, the animals are not forced to live in full view of people. Marieta plans Harnas according to what animals need, not what tourists want. The animals inhabit natural settings with trees, grasses, rocks, and water holes. They hide whenever they want and only show their faces when they hear the truck that brings them food. Sometimes they don't even come for that—although if an animal doesn't show up for several feedings, someone will go into the enclosure to make sure the animal is healthy. Not eating is often a sign that the animal is experiencing a problem.

When they first began rescuing animals, each enclosure was built, rebuilt, redesigned, and moved according to animal needs, as mostly baboons and monkeys poured in—with an occasional rescued cheetah. Then, early in 1987, a call came from a friend in South Africa that would begin the real transformation of Harnas. A zoo in Port Elizabeth was closing, he told them. It was a terrible zoo, to begin with, for any animal to live in, wild or otherwise. As visitors toured the zoo, more and more of them were horrified by how small the cages were, how unclean the surroundings, and how miserable the animals appeared. People in the community became incensed with the conditions and petitioned for its closure, and the government complied under public pressure. The animals would all have to be sent elsewhere or be put down. Marieta and Nick's friend was desperate for their help to save what animals they could.

Without hesitation, Nick and Marieta flew to Port Elizabeth and told the authorities that they would take *all* of the animals. The government of Namibia had different ideas, however, and this is when Marieta and Nick first came up against laws and red tape that often frustrated their attempts to take in animals that needed a home. Namibian law prohibits any non-native animals from entering the country. It's a wise law, actually, in light of the non-native and invasive species that have devastated populations of animals and plants in other places around the world.

Alan Burdick, in his book *Out of Eden*, describes the invasion of the natural world by non-native species. In Guam, for example, the brown tree snake, originally from Australia, arrived, probably in the wheel wells of airplanes, and now there are "more brown snakes . . . per square mile than anywhere else in the world," snakes that have "almost entirely eliminated . . . Guam's native bird population." In the United States, the plant kudzu was introduced from Japan as erosion control, and it now over-runs the South, covering and smothering other plants. No one has been able to find a good use for the unrelenting plant as it continues to proliferate. With problems like these in mind, the government of Namibia tries to prevent this kind of crisis from happening in their country.

This law kept the van der Merwes from taking all the animals, but they did obtain their first four lions, two big baboons, and a porcupine. Because lions in the wild have become increasingly scarce—they have been considered extinct in Namibia since around 1995—Marieta was excited to bring some to Harnas. The lions had not been hand-raised, though, and they couldn't be approached easily. They were overweight and sluggish because their cages had been so cramped, and their fur was matted and mangy. Released into a spacious natural environment, they were confused at first, but just like pets who move to a new house with their owners, the lions only needed time to settle in and explore. They lost the hopeless expressions that had inflamed so many visitors to the Port Elizabeth zoo. They started to develop healthier coats, manes, and eyes.

They remained wary of humans, however, and because the lions didn't have to hunt for their food, they remained portly. Because Marieta wants her animals to feel as wild as possible— freedom within the constraints of an enclosure—adult animals are fed from a distance. Volunteers and Bushman workers feed lions, for example, by throwing chunks of meat over the fence— not by hand-feeding. Such separation from humans makes veterinary work harder, but Marieta wants her animals to feel freer.

Some strict conservationists might say that Marieta's animals are not living a "natural" life, are not in any way "free," and therefore they should be put down or released, even though they lack the skills to survive in the wild. Their argument is that wildlife shouldn't be tamed or contained, that *wild* is the only option, and that if an animal is maimed or genetically impaired, it should be left to die rather than live on "unnaturally." But with the intervention of people, industrialization, and widespread agriculture, "natural" barely exists anymore. A natural life is no longer an option for many animals—especially orphaned, injured, or abandoned ones. For 75 percent of Marieta's animals, the choice is not between *wild* and *free*: the choice is between *alive* and *dead*.

Marieta, along with many conservationists, believes we have already moved past the time when we can let nature correct itself because we have intervened negatively too often and for too long in the lives and habitats of animals. Jane Goodall, chimpanzee expert and conservationist, agrees and writes in her book *Hope for Animals and Their World: How Endangered Species Are Being Rescued from the Brink,* "We have messed things up for so many—it is up to us to put things right."

Today's crisis requires a radical intervention to save endangered species such as the African wild dog—the second most endangered carnivore in Africa, after the Ethiopian wolf. "People must intervene now," Marieta says fervently, "or—because of what we have *already* done to the world—we will lose animal after animal to extinction. These animals are not dying naturally. They are dying because of us!"

The same controversy often comes up in the case of zoos, especially inadequate zoos like the one in Port Elizabeth. Even animals in well-designed zoos are "sacrificed" for the good of their species. Many people who go to Harnas, zoos, or other wildlife centers to see a beautiful lion or tiger or polar bear, make a personal vow—and often a financial gift—to try to save these animals from extinction. These animals are ambassadors for their species as well as for every other species that shares the same environment. People might not feel the same love for a snake or lizard, but that creature benefits by residing where the protected lion lives.

Nico's wife, Melanie, believes strongly in Marieta's philosophy and vision. "It's a common belief in Namibia that a wild animal should be wild. I've talked to die-hard conservationists who believe it's better to put down a three-legged animal rather than let it survive because it's not 'normal,' but my mum says

'*Bullshit.*' She is a very strong pro-life animal worker who says every life has a right to go on. A three-legged wild dog *can* enjoy life and find happiness."

With survival of a species that is headed for extinction on her mind, Marieta observed closely her first group of lions. After they settled in and became more content with their new surroundings, the idea of breeding this endangered animal came easily to Marieta. Once she had established the four lions in a large enclosure, a man came out from Gobabis to see the lions. He had his own wildlife farm, where he bred lions, but his purpose was to offer the lions to hunters, who often shot them in a small enclosure just for the trophy head to take home and mount on their walls. He told the van der Merwes that he had too many lions right then, and he offered some for sale. Marieta jumped on the idea, bought four more lions—young ones—and began her breeding program.

These new lions were relatively wild, so Marieta had no hands-on work to do. She simply had to put the right animals together and hope for cubs. But the lions from the zoo were so fat when they arrived at Harnas that it was impossible to tell whether they were pregnant, and Marieta couldn't get close enough to do an examination. In the case of a lion, darting the creature—even just once—will make that lion suspicious of people for the rest of his life. Marieta saves darting for life-and-death situations only.

On Christmas morning, 1987, Nico went out for an early walk to check on the lions and feed them. He saw something yellow and tawny on his side of the fence. Scooping it up in his arms, he ran home to his mother.

"What is that?" she asked, and Nico shouted, "There are baby lions out there!"

Christmastime can be sweltering in Namibia, the beginning of summer in the savannah, and when they ran outside, it was so hot that the scorched dirt burned their feet. At the fence where the cub had been found, Marieta saw something that made her heart ache: two cubs were dead and another was dying. The mother hadn't taken care of the newborns, hadn't even put them in the shade, and the two had died from heatstroke. By the time they carried the third cub back to the house, it, too, had died. The only reason Nico's cub, a female, had survived is that she had found an opening under the fence, then crawled to the road where Nico found her.

When Marieta and Nico returned to the house, Nick was waiting for them, wondering where they had gone.

"What have you got?"

"It's a lion cub!"

"Why is it here and not with its mother?"

"The mother didn't want it," she explained, "so now it is *my* baby."

"Oh no!" Nick roared. "No! No! This is too much! I will not have a lion in our house!"

But Nick's mandate fell on deaf ears. Marieta had already decided on a name, Elsa, after the famous lioness in *Born Free*, and in her mind she had taken on the role of mother for the cub.

At first Elsa had a hard time with the bottle feedings Marieta gave her three times a day. Mostly she threw up the food—and once she had a convulsion. Local vets were no help, because raising a lion cub from birth was virtually unheard of. Marieta feared that Elsa would die, but then she and Nick talked to a man in South Africa who raised lions. He told them that lion cubs need something in their stomachs all the time, so Marieta must feed Elsa hourly. At other times Elsa should suck on a pacifier. Although this arrangement meant sleep deprivation for Marieta, she gladly sacrificed sleep for her new baby. Elsa began to grow and thrive, turning into a rambunctious, tawny, mischievous kitten.

Like all kittens and puppies, Elsa showed enormous curiosity, climbing into every shelf and cupboard in the house and ravaging all organized workspaces. Part of the problem was that Elsa lacked siblings and she needed playmates. Again using the instinct that has served her so well, Marieta introduced a beagle named Serabi to eat, sleep, and play with the baby lion. The two animals became close friends and went everywhere together. The dog acted as a maternal guide and watchful eye for the lion cub—even though the lion was so much larger and stronger right from the beginning. In the animal kingdom size is less important than attitude and an established hierarchy. As Elsa grew and grew, she still showed deference to this 25-pound dog. The relationship between Elsa and Serabi set a precedent for all future lions, leopards, and cheetahs on Harnas: all of them would be assigned a dog companion to raise and nurture them.

Elsa lived with the family for two full years—until she had reached nearly her full growth. Adult female lions weigh anywhere from 250 to 400 pounds (males reach 400 to 500 pounds), so having a pet around the house that size was eventful, to say the least. Elsa loved her "pride" of people and treated them the way she would a pride of lions, including practicing the stealthy moves of a cat on the prowl to prepare herself for adult hunting.

Marieta recalls, "Elsa had a habit of sneaking up on people and pouncing on their back, trying to bring them down to the ground. At first we all loved the silent attacks that sent us into fits of giggles. Eventually, though, as Elsa got bigger and bigger, it got more dangerous—even though we had taught Elsa to retract her front claws when she attacked. First these attacks were cute, then a nuisance, and finally a real danger. But someone came up with the idea to put a bell around her neck so we would always know where Elsa was. No more surprise attacks from behind!"

Elsa slept each night in Nick and Marieta's bed. She'd settle in between them, taking up more than her share of space. Her belly exposed and her head thrown back, Elsa snored gently in rhythm with Nick. Like any responsible member of a household, Elsa refrained from urinating in inappropriate places—including, of course, the bed. Instead, she nudged Marieta with her cold nose when she needed to be let out.

One such night Marieta forced her eyelids half open, realized her responsibility, and groaned as she got out of bed and went to the front door, opening it just wide enough for Elsa to get out. Elsa followed obediently and disappeared into the garden area. Once Elsa was outside, Marieta went back to bed, still half asleep, leaving the door ajar so the lioness could come back in. After a few minutes, Elsa did return, padded into the bedroom and jumped exuberantly back between her two sleeping partners. The pounce was so powerful and Elsa was so big that Marieta and Nick both were thrown outward and off the bed, like children bouncing off a trampoline.

The acrobatic wake-up call at first surprised and scared the couple, but once they crawled up from opposite sides of the bed, saw each other across the mattress, and realized what had happened, they began to laugh, climbing back onto the bed and wrapping their arms around Elsa from both sides.

Schalk loved to play rugby and soccer with her in one of the open areas near the house. Elsa picked up the games easily and became a true competitor. She would never take out her claws but played fair in trying to take the ball away from Schalk. Other times, she prowled the courtyard and garden, playing with Serabi and the other dogs. She also loved cars and trucks—especially climbing up and sleeping on top of them. Less sturdy vehicles ended up with a concave roof as Elsa grew, and visitors were warned not to park close to the house, where Elsa prowled freely.

Even though people were told of her destructive habit, some continued to disbelieve the stories. The local schoolmaster visited one day and parked his brand-new car near the house. Marieta warned him to move it, but he claimed he'd only be a few minutes. An hour later he returned to his car and found that Elsa had had her fun with it, collapsing the roof, hood, and trunk, removing the rearview mirrors, chewing on the tires, and nearly dismantling the windshield.

Eventually, to protect other cars, Elsa was given a broken-down car for her own enjoyment. She quickly made it her own, sleeping and climbing into and on top of it. Sometimes she even convinced her caretakers to feed her inside it; like a customer parking at a drive-in restaurant, she sat in the front seat awaiting her meal.

Because Elsa seemed like a lively child most of the time, Marieta could almost forget that she was a wild animal. Elsa's instincts were still intact, and by nature she was a hunter and killer. One time she killed four dogs that were protecting the sheep—probably because the dogs barked nonstop whenever they saw Elsa, irritating and angering her. The deadly strike seemed to be a revenge killing, lion style. The family, especially Schalk, was stunned by this attack. Schalk was angry at Elsa for a long time until he realized that Elsa was just a lion, doing what a lion does.

He had been expecting her to act like a human, and that was his mistake—not hers. The dogs were tormenting her, and the wild lion within her emerged, despite her domestic upbringing. Another time she got out of her fenced-in area and killed four young oxen—with one blow apiece to their necks. Life and death on Harnas are closely linked, and each encounter with wild animals taught the van der Merwes a new lesson.

On the other hand, Elsa nurtured and protected the newborn puppies of her dog companions, growling at any creature that came too near. She also saved other animals—mongooses, meerkats, birds—when they were threatened by predators. She seemed to know exactly who was in her pride. Discipline eventually came in the form of a wheelbarrow. For some reason, the large animals on Harnas were terrified of wheelbarrows, and the family soon learned to keep them handy—placed all over the farm—in case they needed to control Elsa or other animals that wandered free. When visitors arrived and parked their cars, a wheelbarrow was waiting for them to push on their way to the house as a protection against any large free-roaming creature.

Elsa also had a brief film career, "starring" in several South African productions. She was generally cooperative and seemed to love "acting." She followed the director's instructions—yawning, growling, and pouncing—right on cue, with the help of at least one family member who always went with Elsa. One film, though, proved to be especially challenging for the young lion, and she refused to do what the director wanted. She seemed distracted and even depressed. She wouldn't eat, wouldn't sleep, and just lay on the ground. Nick finally realized why—they had left Serabi at home. Elsa missed her companion, and she wouldn't work because she was lonely. In desperation, Nick called home to a neighbor and paid him to go to Harnas, get Serabi, and put him on a plane to South Africa. Once the dog arrived, Elsa was overjoyed to see her buddy and began cooperating with the director completely.

On another movie location near Jeffrey's Bay, South Africa, Schalk went with Elsa to take care of her and make sure she did her job. Schalk was only 18 years old, but was strong and mature enough to handle her by himself. As they drove the approximately 500 miles together in an old Ford truck, Elsa sat in the front seat with Schalk. His life was never dull on the roads, at service stations, or restaurants. Whenever people noticed the rather hairy blonde in the passenger seat and then realized what she truly was, they were startled and sometimes came close to crashing their vehicles.

In Jeffrey's Bay, the production company put them up in an old farmhouse on the outskirts of town, but apparently the company forgot to tell everyone that a lion would be staying there. One morning four Cape Town workers drove a truck to the farm to deliver a portable toilet to Schalk, because there was no bathroom in the old house. The men were unloading the toilet when Schalk, who never quite understood the ruckus that was made over one lion, came casually down the farmhouse stairs, Elsa strolling beside him. The shocked and frightened men saw Elsa, screamed in unison, dropped the toilet, and all piled into the front cab of the truck, sitting on top of each other, rolled up the windows, and drove away as quickly as possible. Schalk and Elsa looked at each other in surprise, as if to say, "What's the fuss?"

Schalk remembers this time in Jeffrey's Bay with Elsa as idyllic. He and Elsa slept together on a mattress on the floor each night and spent their free time strolling through the pastures, playing on the beach, and sharing what he calls "quality time" with his favorite girl.

After two years, though, the family made the hard decision that Elsa was too big to remain a household pet, especially because she had come into heat, something female lions do every two weeks or so until they mate. Anyone who has had a house cat in heat knows the annoyance and distraction that can be. In hopes of breeding her, the family moved Elsa out to an enclosure with a male. Her first mate didn't work out because he contracted blood poisoning and died soon after Elsa joined him. But eventually Elsa did breed with another lion—Schabu—who had come to Harnas about a year after Elsa did, and she lived a contented, serene life on Harnas, intermixed occasionally with film work and other public appearances.

Schabu had also been hand-raised and lived as a member of the family for a time. Schabu was very attached to Max, who worked for years as general farm manager and master of many trades. Wherever Max went, Schabu followed, and Max often showed off for friends and visitors by wrestling with Schabu. One day, Marieta and Nick were shopping in Windhoek, a three-hour drive away, and Max was sitting in Marieta's kitchen with Schabu. Suddenly Max heard something strange outside and left the house to investigate. During his absence, Schabu paced the house, becoming more and more frenzied, worried that he had been left behind. Finally, in his desperation to find and be with Max, Schabu jumped right through the living room window, shattering it into a thousand pieces.

When Max saw what had happened, his first thought was for Schabu's safety, but he checked the lion over thoroughly and found he was uninjured. Next, he knew that the window had to be fixed, and since Nick and Marieta were in Windhoek, he called there to ask them to buy the new pane of glass. This was in the days before cell phones, but Windhoek was still a fairly small town where people knew each other, so Max called a store that he knew Nick and Marieta frequented and left a message: "Schabu jumped through the window. Call home." When the store manager asked Max who Schabu was, Max answered: "A lion."

The message flew through Windhoek like a scandalous rumor. First the message was "A lion jumped through a window on Harnas," but soon grew to "Lions are escaping from Harnas" and then "Lions are jumping through windows at Harnas and attacking people." Finally the rumor reached Nick and Marieta, who discounted it immediately. Nick told bystanders, "This can't be. We don't have wild lions at Harnas!" Yet the rumors continued. When Nick finally got through to Max he learned the truth: "No!" Max informed him. "I just wanted you to bring home a new window because Schabu jumped through it to follow me!"

Schabu also took trips with the family to South Africa, especially since—unlike most lions—Schabu liked to swim. "When we went on vacation one time to the beach in South Africa," Marieta muses, "we decided to take Schabu and his cocker spaniel companion since we were driving. It was easier, in those days, to cross the border with a lion, and we got a permit to bring our furry child with us."

The difficult part was finding somewhere that would take a family that included a lion as a pet. Once they found a suitable place near the beach, the task became convincing the older lady who owned the house to allow them to bring Schabu inside, but eventually she decided it might be the most exciting thing she had ever done, so she agreed. Marieta covered all the furniture in the house with layers of blankets to prevent Schabu from gouging the upholstery, and the family happily settled in.

Once word got out that a lion was living at the house, the windows were always full of people's faces peeking in to see Schabu. The van der Merwes took Schabu with them everywhere they went—playing ball on the beach, cavorting in the waves, chasing with the children along the sand, and sightseeing through town—with Schabu's head peering out the back window of the car or sticking his nose out a side window when they stopped at a light or gasoline station.

At first people were understandably frightened by Schabu, especially when he ran free on the beach. Once they saw how much fun the three children were having with Schabu, though, a whole group of people joined in, bringing home amazing stories of a playful lion on the beach.

Schabu was only about a year old on that trip. As he grew older, his disposition changed, as males often do. Generally he was naughtier as a cub and more dangerous as an adult than Elsa. He was a smart lion who taught himself to open gates from both sides, often leaving his enclosure at night to go hunting, occasionally causing havoc by killing some of Nick's game. Schabu's cocker spaniel was named Sacha. Like Serabi, Sacha kept Schabu out of most trouble and kept him settled down on family trips and while inside the house. On a walk one afternoon with Schalk, though, the three ran into a black mamba— the deadliest snake in Namibia, with a head that is fittingly described as "coffin-shaped." The snake's body can reach up to 12 feet in length, and the creatures move with incredible speed, striking with great accuracy. Schabu wanted to play with the snake, not realizing the danger. Before Schalk could move, Sacha rushed in between Schabu and the snake to protect his lion. The snake struck, killing Sacha quickly, as its venom paralyzed the lungs.

Without his dog companion, Schabu became more jealous of Elsa's closer relationship with the family members. As he turned more aggressive, members of the family had several run-ins with Schabu when they came to see Elsa in the enclosure. One day Nico entered their enclosure, not realizing that Elsa was again in heat. Schabu rushed Nico and bared his teeth, but Nico escaped by jumping over the fence—even though it meant grabbing the electric wires and burning his hands.

Schalk was also attacked once when he went into the enclosure to see Elsa. In a fit of jealousy, Schabu grabbed Schalk's leg with his powerful teeth, shaking him around like a toy. Elsa reacted by attacking her mate, biting him on his backside. In order to respond to Elsa's attack, Schabu had to release Schalk, who then escaped—injured but alive.

Schabu is dead now, but his legacy lives on in many lions on Harnas. Schabu and Elsa are the parents of Sher Khan, Teri, and Savanna, and the grandparents of Lerato, Kublai, and Macho. Elsa is still tame. Today, she lives in an enclosure with Sarah— who never had cubs and is the last of the original four lions saved from the South African zoo. Schalk fondly remembers playing rugby with Elsa and claims he could still go into her enclosure today and feel safe. Marieta often walks out to see Elsa and calls her to the fence. Elsa responds by coming to Marieta and rubbing herself on the fence so Marieta can reach through, give her a good scratching, and talk some soft baby talk that only the two of them understand.

Granddaughter Aviel with Grace

chapter four

BABOONS *in the* BED

*A nation's greatness and moral progress can be judged
by the way it treats its animals.*

—Mahatma Gandhi

SAVING ANIMALS BECAUSE EACH DESERVES TO LIVE IS A
guiding force at Harnas. Many people believe animals have souls,
and if that is true, the soul of a lion is worth saving, as is the
soul of the smallest tortoise. Virtually every animal at Harnas
is being given a second chance and would most likely be dead
if it not for Marieta van der Merwe. She never turns an animal
away just because it's unwanted, it's not in the right place, or
it's not perfect. For this reason, Harnas has been home to many
creatures that have illnesses, special needs, and defects of various
kinds. Every animal has the desire to live and, Marieta would
argue, they have the *right* to live. These animals' stories bear telling because they illustrate the lengths to which Marieta will go to
save wildlife—one animal at a time.

In 1992 the family was visiting Marieta's father on the farm where Marieta grew up, Tennessee. One evening, Marieta, Nick, Marlice, and her friend were sitting on the porch when they saw a Bushman in a donkey cart wheel by with something running after it, tied to the cart with a rope. The animal could not keep up with the speed of the cart and kept falling over, knocking its head, and being dragged along. From its loping run, the animal clearly wasn't a dog, so Marlice and her friend ran after the cart to find out what exactly was going on and what this creature was.

They discovered it was a small male baboon—smaller than he should have been for his age. He was an off-white color rather than black like a normal baboon, so they thought perhaps he was albino (although he turned the normal black/brown later in his life). Marieta wanted to rescue him from his abusive owner. The Bushman wanted money for the baboon, and even though Marieta doesn't like paying money for animals because it creates a market, she made an exception in this case.

They named him Boertjie, pronounced "bore-kee," which means "little farmer." From the beginning Marieta realized that he was not like any other baboon. He had trouble walking and had occasional fits or seizures. Marieta and Nick took him to Windhoek to a trusted veterinarian who did a brain scan on him. Not only did they discover Boertjie had a form of epilepsy— perhaps brought on by repeated blows to the head—but he also had Down syndrome. Just like people with Down syndrome, he had certain physical characteristics: one crease across his palm rather than many, motor-function difficulties, and probably the inability to develop mentally past a certain point. He would always remain a "baby" baboon, even though he would grow and become powerful, with a startlingly strong bite.

The vet suggested putting Boertjie down, saying it would be for the best. But Marieta imagined that if Boertjie could talk, he might say, "But I don't want to die!" She made the decision to keep him and make his life as happy as possible, so the vet gave her pills to control his fits. The pills were expensive, but eventually someone sponsored Boertjie and paid for his medicine. At first the family put Boertjie in with other baboons, but, like animals of many species that sense weakness in another, the other baboons attacked him mercilessly. They were not being vicious, but natural law requires that the strong survive to reproduce, and the baboons were just making sure that Boertjie's genes ended with him. Once Marieta realized what was happening, she removed Boertjie and put him in his own enclosure.

He was never too lonely, though. Most recently, his enclosure was set next to that of the baby baboons, and he was curious and entertained as he watched their antics. Mongooses sometimes visited Boertjie in his enclosure, and he watched them with great curiosity, not seeming to care when they stole an occasional bit of his food. He never attacked one, but let them dig under his fence and come in. They, too, sensed that Boertjie was not a threat. Even the smallest mongoose acted unconcerned that a large primate was watching him intently from only a foot away.

The volunteers who chose to work with Boertjie loved him. Despite his two-inch canines, he was usually as gentle as a child. Still, because Marieta never knew if a seizure might come upon Boertjie, she limited volunteer exposure to "through-the-fence only." His hands were clumsy, so volunteers fed him several times a day, one piece of fruit at a time. He opened his mouth willingly and let them slip each slice of banana or orange or apple between his huge canines and into his mouth.

His feedings took much longer than those of, say, the cheetahs, lions, and other baboons—all of whom gobbled up their food with gusto. The volunteers who fed him knew about his slower eating habits, and they scheduled the extra time. Boertjie enjoyed the human encounter so much, and as a volunteer, there was something soothing about gazing into those deep chocolate eyes, the perfect audience for whatever subject the visitor might want to discuss.

In between meals, volunteers often stopped at his enclosure and scratched his back, neck, and head through the bars, sometimes cracking peanuts—one of his favorite treats—and slipping them between his teeth. He'd lean against the bar willingly, close his eyes, and go into a baboon trance while being scratched and talked to. One volunteer, Kate, showed up every afternoon with a book and read aloud to Boertjie. He sat rapt with attention, and when she stopped to turn the page or take a drink of water, he leaned toward her as if to say, "And then what happened?" With one hand on her book, one hand scratching his back, they were both content.

Another volunteer who had epilepsy herself felt especially close to Boertjie. She claimed she knew exactly how he felt and, without permission, she opened his gate and went in to sit with him. Marieta was soon fetched and she ordered, "Out! You cannot stay there! He could kill you!" But the volunteer refused, and Marieta left, muttering, "I cannot watch. This cannot end well." But it did. The volunteer sat in his cage almost daily, feeding him, scratching him, and talking to him. He never hurt her—never even moved toward her aggressively.

After 16 years at Harnas, in July 2008, Boertjie died, having led the best life he could, cared for and loved by so many. In the newsletter Harnas sends out quarterly, Schalk's wife, Jo, wrote about Boertjie's life and death: "Even though some eyebrows were raised surrounding letting a baboon live with epilepsy, we now know for sure that God had allowed Boertjie's time here on earth for the incredible impact he had on people all through the years. Not only did he make us think, feel, and laugh, but he taught us great lessons in compassion, empathy, understanding, and patience."

Extreme situations like Boertjie's sometimes call for extreme solutions, and Marieta tries to look at each animal separately, in light of its history, any abuse, and any physical problems. As a result, each animal Marieta rescues and adopts can count on receiving the best possible individual care from her.

Baboons star in many of Marieta's rescue stories, because they are so endearing as babies, become so mischievous and destructive as they get older, and eventually grow stronger by far than their owners. Such is the story of Sarah. A couple phoned Harnas and said they had a female baboon that had grown up with their sheep and goats from the time she was a small baby. She lived in the corral with the other animals, but the corral was close to the house, and as she got older, Sarah became curious about what was going on inside. Curiosity is a natural inclination for baboons, but the couple didn't anticipate the intensity of this personality trait. While they were away from home, Sarah often broke into their house and explored—which for a baboon means dismantling things, throwing things on the floor, and breaking objects, not intending to do harm but simply wanting to see how things worked—how they were put together and how they came apart.

Sarah's behavior, which the couple mislabeled "naughty" and "destructive," was driving them crazy, and they finally decided they had to get rid of her. Marieta drove out to the couple's house and met the baboon. She watched carefully while Sarah mingled with her goat and sheep friends. They treated each other as "family," so Marieta worried about taking her from what Sarah considered her *troop*. Marieta told the couple that she would take Sarah but only if she could also take one of the ewes in order to soften the adjustment of relocation. The couple agreed, and Marieta left for home with the baboon and the sheep.

At Harnas, she took the two animals to an outer post and put the sheep with 13 other sheep and four goats that were already in an enclosure. The ewe seemed fine with her new companions. Then she put Sarah inside a smaller enclosure within the sheep and goat area so the Harnas animals could get used to having a baboon living among them. Marieta left them alone for a week, sending someone out only for feedings. At last, she released Sarah among the sheep and goats, and she immediately became friends with all of them. She especially became attached to a goat named Gilbert, a large white male with six-inch crescent-shaped horns.

Sarah decided her purpose in life was to look after the goats and sheep. She took her job seriously, acting like a shepherd protecting her flock. When the grass grew high, she jumped on Gilbert's back and rode him like a horse, sitting up tall and holding his horns as if riding a motorbike. Baboons sometimes sleep all day and play at night, especially when it is hot, so if Sarah grew tired during the day, she lay on Gilbert's back, draping her arms and legs over his sides, sleeping while he carried her around. At night while the animals slept, she guarded them, chasing away predators like jackals or hyenas.

When the goats and sheep had babies, Sarah held them in her arms, looked carefully into their ears, picked off fleas and ticks, and groomed them the way baboons groom each other. Sarah also began nursing from mother goats and stopped eating the porridge brought to her every day by volunteers. The goat milk satisfied her. The goats didn't mind and let Sarah have all the milk she wanted because she was so gentle.

When the goats and sheep slept—in a circle as they do—she sat in the center and held their babies close. When the trucks arrived bringing food, volunteers, or tourists, she ran around frantically trying to protect her flock from the "intruders." She became famous as a tourist attraction because people loved watching her shepherding antics.

One day Marieta and Nick got a phone call saying that four wild lions had crossed the Botswana border, a 90-minute drive away, and were headed toward Harnas. The callers were hunters, and, as usual, they were chasing the lions, hoping to shoot them for trophies. Without understanding the mission of Harnas, the hunters asked to be informed if the lions showed up at the farm, so they could come and kill them. Marieta told them, "Of course I will!" and then laughed as she hung up. She had no intention of allowing the men to step one foot on her property—much less enabling them to shoot one of the most endangered animals in southern Africa.

The next morning revealed that the lions, indeed, had arrived on Harnas property. When the truck went out with food for the sheep and goats, Sarah and Gilbert were gone, and five goats were dead. Lion spoor was everywhere. The lions had dug a hole under the fence and gotten through to the animals. Two hours later, having tracked the lions to Harnas, the hunters arrived with eight hunting dogs. Marieta argued with the hunters, telling them they could not hunt on her property, especially with dogs. The dogs used in lion hunting are usually "sacrifice" dogs that have been bred especially to chase, find, and fight with lions. Such dogs are regularly killed or wounded beyond healing in the fight, but the hunters don't care—as long as the dogs do their job in finding the lions before they die. Marieta ordered the men off her land.

The hunters knew that the lions would probably turn around at some point and head back to their original territory in Botswana, so they headed east again, tracking and driving the animals. Eventually they found and shot one of the lions, but three got away. Lions in the wild have very few places to hide, though, so the fate of the other three was not promising.

Meanwhile, Marieta and Nick were worried about Sarah and Gilbert. They searched for them all day, and the next day as well. On the third day, they heard the unmistakable call, *baaaaa baaaaa*, of a goat. They followed the sound and found Gilbert, hidden under a thorny bush, calling out for help. He was scratched and bleeding from the lions, which had tried to reach him under the bush but couldn't quite stretch far enough through the long thorns.

In front of the bush, in a position of protection, was the slain baboon. Sarah had put Gilbert, her favorite goat, under the thorny bush to protect him, and then she stood guard, attempting to fight off four lions to protect her friend, sacrificing her own life for the animal she had been entrusted with as a shepherd. If Sarah had been a human, she'd be heralded as a hero—and she is at Harnas.

Stories like this eventually made Nick almost as dedicated to saving wildlife as Marieta. When she first started taking in wild creatures, Nick tolerated it mainly because he loved her and wanted her to be happy. But somewhere along the line he became a true believer in Marieta's mission. One of the animals who convinced him was Tommy, a baby baboon Nick came to love. Everywhere Nick walked, Tommy rode on his shoulders. In Nick's truck, Tommy sat in the passenger seat. Marieta remembers one time when Nick stuck his head out of the driver's window, looked back at her, and waved as he drove away. Tommy, at the same time, stuck his head out the passenger window and made the same gesture. "It was something about the symmetry, the similarity of man and beast, maybe just the fun of seeing Nick with his best friend—it made me laugh and laugh as they drove away!"

Every night at six o'clock, Nick shouted, "Come, Tommy! It's time for a shower!" And off the two went for a shower together. Then came another nightly ritual at the dining table. They sat across from each other with a dish of ice cream and a can of Coke between them. Nick took Tommy's spoon, dug some ice cream out, and handed it to Tommy, who ate it lustily. Then Nick took his turn, with his own spoon, and then Tommy again. Back and forth they'd go, each watching the other in between bites. At intervals, each took sips of Coke.

After that they watched TV in bed. Tommy loved the Discovery Channel—except when he spotted something on it that scared him, like loud airplanes or snakes. Tommy would dive beneath the covers and peer out fearfully from between his hairy black fingers.

Just like a human, Tommy was inconsolable when Nick died, grieving for his best friend. To this day, Tommy—who is living with other baboons in a large enclosure close to the guesthouses—will sometimes escape and seek out humans, perhaps looking for a soft bed, a little TV, or a taste of ice cream on a hot day. He seems as much at home sitting on the porch of a guesthouse as he does cavorting with other baboons, hanging on trees and chasing each other around, screaming wildly.

It was Tommy who led the "siege" on my cottage one morning, trapping me for a while, although I didn't know it at the time. I just saw a big male baboon with scary teeth peering down at me through the hole he'd torn in my roof. And it was Tommy who stayed when the others left, sitting on my porch banister, eating my orange. I took pictures of him and showed them to Marieta. She laughed, "Oh, that's just Tommy! He wouldn't have hurt you. He was just paying a social visit."

A few years later, a baboon came to Harnas that soon earned the name of Houdini. He seemed to be able to escape any enclosure at will, and while many baboons regard an escape as a challenge to be enjoyed, Houdini has perfected it to an art. Electrified fences are nothing to this high-flying acrobat and contortionist. Recently he has started leaving not to escape something but to find someone: Mara Kuhn, a preschool teacher hired for Marieta's grandchildren and the Bushmen's kids who attended the Cheeky Cheetah Daycare that Jo had begun. Mara dramatizes the story with a variety of voices and gestures, pretending annoyance at this "bloody bother," but it is clear she feels special because this baboon has fallen in love with her.

In three months, Houdini broke into Mara's room four times. The first time, Mara left her door open while she went to a communal bathroom to take a shower. She was gone only 20 minutes, but when she returned, her room had been reduced to a shambles. Books and papers for her school were scattered everywhere, her lotion had been opened and spilled, a chair was turned over, and the blankets on her bed had been rumpled. A glass vase and porcelain frame lay broken on the rug.

"It was partly my fault because I left the door ajar," Mara admits now, resignedly. "I cleaned up the mess and went to report to Marieta that a baboon—probably the infamous Houdini—was on the loose, but when I returned to my room, I stood at my door with my mouth hanging open. Houdini had been back in just the ten minutes I had been away! The damage wasn't as severe—his time had been limited—but there were more broken things on the floor and my blanket had been dragged outside."

This time Mara felt no resignation, and she wasn't amused. A favorite refrain around Harnas—"that bloody baboon"—could be heard all the way across the courtyard. "She learned to keep her door closed!" Marieta laughs.

Time makes us careless, though, and a month later, Mara left her window open just a few inches, though locked with a special latch. Baboons have hands like people, though, and they're famous for being able to unlock anything. Gates to baboon enclosures on Harnas always have two locks. When Mara came home for lunch, her front door was closed, but when she walked in, Houdini was sitting on her bed, examining her alarm clock. He looked up at her innocently and greeted her with a "Uunnhh, uunnhh." She could have sworn he was smiling. "Nothing was broken this time," she sighs, "but what do you do when you find a baboon on your bed?"

The fourth time pushed Mara over the edge. Normally calm and dignified with a preschool-teacher voice, Mara was enraged. She had left her room for only a few minutes to run some clothes to the laundry, and when she got back, her window had again been jimmied and her room was a mess. She ran to Marieta again.

"And he's in there still!" she wailed.

Marieta erupted in streams of Afrikaans as she stomped across the courtyard to the small apartment, with Mara in tow. When they looked in, Houdini was nowhere to be found, so they entered cautiously. Mara started picking up pieces of clothing and other personal items when they heard "Unnnhh, unnnhh" coming from the closet. Marieta approached slowly and eased open the door. At first it was too dark to see much other than the higher shelves that held folded stacks of shirts and shorts. Underneath the shelves, though, a ghostly shirt seemed to be floating in air. Then they both saw the dark hairy hands at the corners holding it up. The shirt dropped and Houdini popped out from behind it, enjoying the game immensely. He opened his mouth wide and laughed, "Aaagghh, aaagghh, aaagghh."

"It was too much!" Marieta laughed and took Houdini by the hand, pulling him up on her hip and walking out to the courtyard, where a Bushman was waiting to return Houdini to his baboon community. Even Mara has to smile now, having gained some perspective on Houdini's love games. She suspects she was supposed to feel flattered to be the object of this romantic attention.

Other primates—both from within Namibia and without—live at Harnas, some victims of purposeful cruelty and some who have accidental injuries. In one cage near the garden area lives the shy Mister Nielson, a South American black-capped squirrel monkey who was confiscated, along with three others like him, because, in Namibia, it is illegal to keep exotic animals from outside the country. This tiny bright yellow black-headed monkey would have been put down if Harnas had not taken him. His three companions, unfortunately, died of tuberculosis, contracted from a Bushman, but Mr. Nielson is especially hardy. At least eight years old, he is only seven inches tall and weighs less than two pounds. Although he is wary of strangers, anyone sitting in his enclosure in the afternoon reading a book or writing in a journal will create a bond with him. Once he feels comfortable, he will often jump down from his perch onto a shoulder and accept a peanut or two as a token of friendship.

Animals, Marieta would argue, deserve the same concern with the same dedication to healing as humans, no matter how overwhelming the undertaking might seem. One story that illustrates Marieta's belief concerns two lion brothers, Sam and Robert.

In 1990, Harnas witnessed a huge number of lion cubs—13 in all—born to the original group of lions from the Port Elizabeth zoo. The family decided they could not keep them all, so they arranged to send the cubs to a lion farm in South Africa. Before they could be taken across the border, though, they had to be tested for a variety of transmissible diseases. Two of the lions, Sam and Robert, were diagnosed with feline immunodeficiency virus (FIV), similar in nature to HIV/AIDS in humans—except that it can be transmitted through saliva and biting, which lions often do to each other during play. It also resembles feline leukemia, which small domestic cats contract. Unfortunately, FIV infects and kills many wild lions in Namibia and South Africa.

How the two cubs contracted FIV is a mystery, although one theory is that when Sam was treated by a local veterinarian for an illness, he might have been infected with FIV through contact with another lion or by an unsterile needle. Sam might have given the infection to his brother Robert while wrestling.

The two young lions stayed behind at Harnas while the other cubs were transported to South Africa. Euthanizing these two brothers was the easier choice, considering the lack of information available about this disease in lions and the expense required to keep them healthy. At the least, they needed a separate enclosure so they couldn't infect any of the other lions on Harnas. By living separately, they wouldn't have to fight for food or territory, nor would they have to expend energy hunting. And by monitoring their health closely, Harnas tried to ensure Robert and Sam the best chance possible to live normal, healthy lives.

Robert survived until 2006—16 years—and Sam is still alive, despite his FIV. Many observers have suggested to Marieta that caring for such potentially sick animals isn't worth her effort. But saving these animals is only part of her journey. Those who work to save animals like Sam learn from his fierce determination to survive. Every visitor to Harnas hears the story of Sam and Robert, and they leave knowing more about the hazards in the wild that are contributing to the extinction of these proud creatures.

But most of all, Marieta knows that Sam is glad to be alive. He has someone who was willing to give him a second chance. When the lions roar at sunrise and sunset, marking their territory and telling the other lions, "This place is mine. Don't fool with me!" Sam is one of the first to roar. And his roar is impressive. He comes directly to the fence, faces off against any human or animal on the other side, and lets go a reverberating sound that you feel in your chest, in your lungs, and in the beat of your heart. He has a roar that tells the world that despite his diagnosis, he is not to be underestimated.

Melanie knows that seeing suffering animals is hard for everyone, but especially for Marieta. "If something suffers, then yes, as a last resort, we will put it down," explains Melanie. "But you can

develop compassion and love in people through these animals—
animals like Boertjie, like Mr. Nielson, even Houdini—teaching them through their encounters with these creatures. They
become ambassadors for all animals. My mum has always pulled
for the underdog, in life, sports, and in nature, too. When she
watches a rugby match, she'll cheer for whichever team is behind,
and if that team starts winning, she'll start cheering for the other
side. It's all about nurturing to her, about change and growth,
and she believes people have the ability to learn to love other
creatures. It's her passion."

Schalk with Zion

chapter five

GROWING UP *in the* WILD

The animals of the world exist for their own reasons.
They were not made for humans any more than black people
were made for white, or women created for men.

—Alice Walker

AS THE ANIMAL POPULATION ON HARNAS WAS EVOLV-
ing, so was Marieta's family. Nico, the one brunette and brown-
eyed child, was becoming a young man with diverse interests. He
loved and respected animals but he did not care to play with them
as much as observe them. Instead, he favored working with com-
puters, building things, and reading books. Marieta remembers
him as the silent, solitary child. Even when he went horseback
riding, he usually went by himself rather than with a companion.
When he was a child, Harnas was a small family endeavor, but
by the time its fame had spread—from exposure on such African
television shows like *Carte Blanche*—Nico had gone to college.

Therefore, Nico missed the relative fame that would shine
on the van der Merwes. People knew that Marieta and Nick

had an older son, but he didn't share the limelight. And even though people saw him as the studious child, he was athletic, too, running track at school, doing karate, and playing rugby. He preferred individual sports to team sports because he felt that a player could hide on a team, whereas an individual either won or lost by his own efforts. All of Marieta and Nick's children were athletic, and all were attractive and popular at school, though usually they were children of few words.

The third child and only daughter, Marlice, was the star, the actress who posed whenever someone needed a photo with an animal. She inherited her mother's beauty and became a lovely girl with long blond hair; a wide, white smile; and hazel eyes. It's no wonder the camera loves her. Her face can be found in picture after picture with animals at Harnas, and she has enjoyed playing the model with all her wild-animal friends because she grew up around them. Even though Marlice sees herself as a "primate woman" like Marieta, drawn to vervet monkeys and baboons, photographers nearly always choose to shoot her with lions. The contrast of the attractive and frail (compared to the lions) female next to the ferocious (but relatively tame) lions always makes for a spectacular shot of Beauty and the Beast.

Besides the animals, Marlice's playmates were the San children, and because of this, she learned the extremely difficult language of this tribe of Bushman. She is one of the only white people in Namibia who is fluent in San—and when she went to school, she knew the complicated language better than she knew the Afrikaans used in the classroom. Her teachers were first puzzled and then angry, reprimanding Marieta and Nick for what they considered a gross oversight on their part. They didn't realize the enormity of the accomplishment this young girl had managed. The language she learned, unfortunately, belonged to a people who live without respect in their own country.

Marlice could often be found with the families of the Bushmen, eating, riding horses, and playing the traditional game of stones and sand. She smiles as she remembers: "The San girls and I also played Crooks and Cowboys with Schalk and the boys. We tracked each other through the bush barefoot!" The children ran through the bush, encountering snakes, scorpions, and other dangerous creatures. They knew what to do and what to avoid, and their parents didn't worry about them any more than if they had been playing at a friend's house in the suburbs of North America. This kind of play as children also helped train them to track animals that were injured or lost. Many times the children were employed to help find these animals and capture them so they could be saved from hunters and poachers, as well as receive the veterinary care they needed.

"When I got to hostel," Marlice shakes her head, "I was baffled by the play of the other white girls. I whined to my mother that I didn't know how to play these stupid games they played. I begged to come home to my real friends."

Marlice recognized for the first time that her childhood was unusual. "When I went to school was when I realized my life was different than others," she says. What each of us grows up with seems normal, and for Marlice, "normal" meant being surrounded by animals of all kinds. She laughs, "My friends probably thought I was a crazy liar with all my stories of cheetahs, baboons, and lions. Not until my best friend from school, Salomi, came home with me for a weekend did anybody believe it. 'It's all true!' Salomi told everyone on Monday, and then they all looked at me differently!"

Nick and Marieta raised all three children not to drink, smoke, take drugs, or have underage sex. Perhaps because these normally taboo issues were openly addressed in their home while they were growing up, the three never felt the need to participate in them. Marlice was offered a cigarette by her mother one day, and together they smoked, choked, coughed, and even vomited. Marlice never had the urge to smoke again. And remembering her own nonexistent lessons about the female body, Marieta made sure that her own daughter never suffered from lack of knowledge. Marlice got all the details about menstruation and sex early on in life—while her friends in hostel got confusing metaphors about bees and pollination.

Marieta made sure her daughter grew up as she had—tough—and this toughness was tested in sports. She often played for school teams in grades above her. In grade school, she played for the high school team in netball, similar to women's basketball, ran both sprints and hurdles in track, and did karate. Marlice played for the Namibian national netball team for four seasons and later competed on the all-African netball team. Like her brothers, Marlice was stubborn, competitive, and determined to be the best, traits they inherited from their mother. They believed nothing short of winning was acceptable, as opposed to Nick's philosophy of life: peace at the expense of compromise.

Marlice and Schalk did inherit one trait from their father, though—dyslexia. In the 1980s in Namibia, this reading problem hadn't been identified and defined, so both children struggled with their schoolwork. The struggle turned them from books to sports, an arena where they could excel and gain self-confidence.

Schalk was generally regarded as the creative, adventurous child, and he has many characteristics attributed to the middle child of any family, such as avoiding confrontation and being independent and secretive. "From the time I was five years old," he says, "I'd leave the house, always barefoot, and spend the whole day with Gou (my Bushman friend) in the bush and not come home till evening. We'd each take a small bag with water, a knife, and a few other survival items and spend the day hunting small birds, building fires to cook our catch, and exploring the terrain. I don't think my mother ever worried about me."

Marieta agrees, describing Schalk as a survivor from the beginning, despite his daredevil antics. When Schalk was nine the family bought a television set, and he loved watching Western movies and kung fu flicks. He and Gou would reenact cowboy and karate movies they had seen on television, and often other children were the victims of Schalk's re-creations. Once, Gou was almost hanged in a scene they were playing from a Western. Another time, Schalk pushed Marlice off the roof without putting any padding below just to see what would happen, and she broke her arm.

He had his own share of mishaps, too. His parents gave him a small Italian motorbike, and he built some ramps to jump over obstacles—as he had seen on television. Once he missed the ramp entirely and smashed into a tree, creating a huge dent in his helmet, not his skull. Instead of crying over his aching head, Schalk cried over his broken motorbike, yelling at it, "Get up! Get up!" and kicking dirt over it as if it were a dead animal.

In contrast, Nico loved school and got high grades and academic prizes. At parent/teacher conferences, every teacher would say the same thing: "Nico is a dreamer. He doesn't pay attention and looks out the window all day. But when he's asked a question, he always knows the answer. And when he takes an exam, he always does extremely well." So his parents never worried about his "dreamy" side. They knew he would excel.

Schalk, on the other hand, didn't want to be indoors in class. He wanted to be out, doing something active. He was a natural at sports, often beating boys much older in boxing, karate, and track. Eventually he played rugby for the national Namibian team in the World Cup for many years, once being named the Most Valuable Player on the Namibian team. Marieta describes him as lacking natural aggression, though—needing motivation to do his best.

At the beginning of a game, he would look a bit lazy and Marieta remembers trying to motivate him from the sidelines: "'Schalk! You play like shit!' and then he would get angry at me and yell back, pleading with me to stop embarrassing him: 'Ma! Shut up, Ma!' And I would yell back, 'But you play like shit!' Then he would make a noise like a cross between a lion and a big baboon: 'Uunnhh! Uunnhh! Uunnhh!' His father would say, 'Now he'll start playing.' And he would. He'd run over everyone—because he must be angry to play his best." She laughs heartily at the memory, shaking her head and laughing all over again.

When the three children had to go away to school this was extremely hard on Marieta, even though she refused to let her children see how it broke her heart to send them away from Harnas from Monday to Friday. The children also were reluctant to leave their home, their parents, and the wild animals that were like family. Other children simply didn't understand them and vice versa.

The first child, of course, was the hardest to let go. At age six, Nico left for a school that was 120 miles away because it was a school located on a farm, and Marieta and Nick thought it would feel more like home, but he cried every Sunday when he had to go back. By the second grade, his parents decided going so far away was too hard on Nico, so they moved him closer to them, to a school in Gobabis where there were fewer children in a room, and so he was a bit happier.

When Schalk went to school, Marieta and Nick expected the change would be easier because Nico was already there, and he could watch over his little brother, but this wasn't the case. School was a miserable experience for both Schalk and Marlice. They were both so tied to the animals and the farm that adapting to school was especially difficult.

By the time Schalk was a few years along in school he had run away three times, once getting as far as the neighboring town of Drimiopses. His teacher contacted Nick, and the two of them met halfway, borrowing a truck Schalk wouldn't recognize so they could sneak up on him. They found him, pulled him into the car, and returned him to school. Schalk claims now that he didn't mind so much about being away from home, but he disliked struggling in his classes, probably in large part because of his reading disability. He would feel unprepared when a test was coming up and think, "Well, I'm going to fail out of school, so I might as well get it over with and leave now." He can laugh about it in hindsight. "School was just not for me," he concludes.

Marlice went to hostel full-time the year after Schalk, but because she lived in the girls' dormitory, she was separated from her brothers a great deal of the time. Her room was near the chicken coop, and when she heard the roosters crow every morning, she was reminded of Harnas, and she started each day crying. Eventually, her hostel *ouma,* or grandmother (who later turned out to be her husband's real grandmother!), took a special interest in Marlice and realized what was happening with the roosters. After that, every morning she slipped into Marlice's room before dawn and carried the small girl to her own room and woke her slowly, feeding her toast, and making her smile before the rooster crowed.

Despite all of the anxiety about boarding school, as adults Marieta's children agree that living away from home was good for them, that it made them tough and independent, and that their children should experience the same. Raising children in a country like Namibia, where only the strong survive, takes a special kind of determination. Namibia is roughly the size of Texas and Louisiana combined, yet it has a population of only about two million people. Although the literacy rate of adults is high—85 percent—most people lack the training and skills to find a well-paid job. Around 88 percent of the people in Namibia belong to one of the ethnic tribes, and most lead a life of day-to-day worry about subsistence, trying to make a living in agriculture. As in most sub-Saharan countries, HIV/AIDS is a major problem in Namibia, with almost 20 percent of the population infected. And although the country has great wealth in mining—diamonds, uranium, copper, gold, lead, zinc, silver, and semiprecious gemstones—and its fishing grounds are among the richest in the world, the income distribution is one of the most inequitable on the African continent.

White people have had the advantage, of course, having descended from the explorers, traders, missionaries, and hunters that came mainly from Germany and the United Kingdom, taking the best land for themselves and using the tribal people as workers—both paid and unpaid. But even with the advantages of education, steady nutrition, and a stable home, life in the rural areas of Namibia can be dangerous and demanding. As a mother, Marieta knew from her own upbringing that a parent sometimes had to be tougher than he or she would like in order to raise resilient kids who can survive the difficulties of living on a farm in wild Africa. Schalk remembers one particular time when his mother had to be especially hard on him:

"I was about five years old, and I loved horseback riding. My horse had only one eye—she had been beaten by the Bushman who owned her before. But I loved her—and we did a lot of things together. One day my father brought me a cowboy saddle—I rode bareback before. So I had my clothes and everything to be a cowboy. But the saddle was brand new and the Bushman who tied it didn't tie it tight enough. As I rode, it slipped to the side and I was riding on the side of the horse, and then suddenly I found myself under the horse!

"I fell off and the horse ran over me. I rolled into some bushes and had scratches all over me from the burrs and thorns. I was crying but my mother told me to get back on the horse and ride again. I didn't want to and I kept crying, but she kept saying 'Get on! Get on!' Eventually, I got on. I was afraid and nervous for the first minute, but then I was fine and was never afraid of horses again."

Marieta also remembers the incident clearly and admits she's "a hard woman." She learned from her father: "If you hurt yourself, you have to try whatever you were doing again or you'll always be afraid. I made Schalk get back on that horse, and then I hit the horse to make it run. He held on and from then on wasn't afraid. That's how you have to survive here."

Marlice believes, though, that her mother is as tough on herself as she is on others, and what Marieta hates to see most in others is self-pity. She never wanted her children to be weak, a lesson she learned from her father. In retrospect, Marlice credits her mother's determination to push her children as having made Marlice the strong and independent woman she is.

Now that Marieta deals with young volunteers from all over the world who haven't had this kind of tough training, she tries to get this lesson across to them, too. "If a baboon bites you," she says by way of an example, "you'll be afraid of baboons unless you get back into the enclosure again." Frikkie's advice is even more surprising: "If a baboon bites me, I bite him back—and he'll never bite me again."

IN 1987, THE VAN DER MERWES STILL RAISED CATTLE, and Marieta's animal rescue was considered her hobby. Because technically Nick was working for his father, when his father bought a butchery in the coastal resort town of Swakopmund, he asked Nick to move there and run the new enterprise. Nick and Marieta made the difficult decision to move almost 400 miles away, leaving most of Marieta's animals on Harnas under the care of her cousin Rudy. The move was heartbreaking for

Marieta and the children, who had to leave their home and many of their favorite animals. Marieta couldn't leave all of them, however, even temporarily, as Nick promised the move would be, and so she brought along some cheetahs, baboons, monkeys, and meerkats.

Two years later, when Nick's father died of leukemia, Nick received his inheritance: six farms, around 114,000 acres, and all the cattle, equipment, and buildings on the properties. Nick and Marieta, who had her own family's inheritance, became one of the ten richest couples in Namibia, although the farms and equipment were in poor condition because Nick's father had been too sick at the end to care for them properly. The family moved back to Harnas and began to figure out this new life that included more income but also much more responsibility. Nick and Marieta hired a manager for the butchery, but within months they closed it down, moving much of the equipment to Harnas where it was put to good use in cutting up meat for Marieta's carnivores.

Life was good for the van der Merwes for a time, although having to renovate the newly gained farms stretched their cash flow. They had the wherewithal to take care of everyone— and every creature—on their land, and Marieta's "hobby" was going well and making her extremely happy.

Then tragedy struck.

In August 1989, Nick flew with his pilot friend Alwyn Bierman to Windhoek to pick up Nico from a school activity. With distances so vast in Africa, planes are often the vehicle of choice instead of spending endless hours in a car bouncing over dirt roads. Many people have airstrips on their properties, as people fly in and out using small planes maintained by their owner-pilots. At Harnas the airstrip was used for bringing in a veterinarian, for guests who preferred to fly, and for emergencies—and occasionally, as in this case, for bringing a child home from school.

The plane was a Piper Cherokee 180, a model one news article called "hopelessly underpowered and, although it is a four-seater aircraft, it must be treated with meticulous care when more than two occupants and their luggage are on board." Bierman was a qualified pilot, and although the plane was filled with three passengers and lots of luggage, he felt confident. As the plane approached Harnas, it passed low over the farm to draw attention to its arrival, as planes usually did, and alert Marieta that her husband and son were home.

Marieta was cutting meat when she heard the plane, and she left the house, running toward the airstrip. The plane climbed to bank, took a right-hand turn, and then, according to witnesses, "stood on its right-hand wing, dropped and seemed to 'flutter' for a moment, making a steep descent." The pilot couldn't regain altitude, and the plane skimmed over the bush, disappeared behind some trees, and exploded into flames. A small mushroom cloud rose into the air.

Marieta—along with everyone else—sprinted toward the wreck. Marlice, out horseback riding, urged her horse into a gallop toward home. Schalk raced toward the airstrip. Marieta ran straight toward the wreck, through thorn bushes and trees, and arrived with bloody scratches on her arms and legs. The fuel tank had exploded, and the plane was a fireball. No one could get near it, and it was impossible to see if anyone was in or near the plane. Pieces of the aircraft were scattered and burning.

The first to emerge from the chaos was the pilot Bierman, who had been thrown from the plane on impact. His hands were burned and his forehead was cut, but otherwise he had escaped serious injury. Suddenly, Marieta heard the word "Ma!" and looked toward the call, but she saw someone she didn't even recognize. Nico's face was blackened like charcoal, and his burned hands, knees, and ankles looked like raw meat. His left arm and hand seemed even worse, blackened and shriveled, and strips and tatters of blackened skin hung from his body. Marieta couldn't even hug him because his injuries were so severe.

He led her to an unconscious Nick, who lay 15 yards away from the wreckage. Nico had dragged him there, saving him from certain death. Part of Nick was still on fire, so three Bushmen, who had grabbed shovels and come running when they saw the crash, began tossing dirt on him, dousing the fire and saving his flesh from further damage. All of Nick's clothes had been burned off—with only pieces of his socks remaining. Like his son, his skin hung in black strips, with the red of his flesh underneath exposed. His ears had been burned into shriveled masses. Father and son were in a state of shock. Marieta's cousin Rudy loaded the injured men into his truck and they all headed to the nearest hospital in Gobabis, an hour and a half away. All the way there, the sensation of burning increased for the victims, and Nico begged his mother to blow on his poor burning hands. Other times he held them out the window to let the wind cool them and dampen the excruciating pain.

The Gobabis hospital was not prepared for such an emergency. All they could do was give each victim a large dose of morphine, dress the worst of the wounds, and transport them to Windhoek—another hour and a half away. In the capital city, Nick and Nico were placed in baths to peel away the dressings and what was left of their clothing. This painful procedure was repeated each day when the dressings were replaced, and neither father nor son could stop himself from screaming. Their agonized cries could be heard all over the hospital.

The pain was relieved only by morphine. Nick stayed in intensive care for over a week, and then demanded to be put in the same room as his son, believing they could bolster each other's strength and encourage each other when they felt hopeless. In total, the men remained in the hospital for five months, receiving care, medicine, and skin grafts. Despite the pain, the screaming, and the constant fear of infection and death, the two maintained a good attitude and sense of humor. Photos of the two show them completely swathed in bandages, their faces like charred meat, especially on the left side. But they're almost always smiling—when their faces could finally permit that muscle movement. Both men radiated strength and determination, and they never showed doubt to others, no matter how worried they may have felt.

Initially both Nick and Nico improved, but after a few months, Marieta noticed that Nick was getting worse. She thought the hospital in Windhoek was not as clean as it could be, and she suspected her husband was suffering from blood poisoning. She noticed that although visitors to the hospital were made to wear surgical masks, ironically, the cleaning people used the same dirty mops on the toilets and then on the floors—never changing them from room to room. Who knew what germs were being spread from patient to patient? When she requested that her husband and son be transferred to a better hospital in South Africa, her doctor became incensed and said, "If you ever have another problem, don't come to me!" Marieta didn't let the doctor intimidate her, though, and she pursued the transfer of her husband and son.

When the hospital in Windhoek still refused to relocate the patients, Marieta became more worried. She could see clearly that Nick was getting worse and she was afraid he was dying. In her independent fashion, obtaining the help of a friend who had an airplane, she arranged the transfer herself. By the time they reached the South African hospital, the doctors gave Nick only a 4 percent chance of survival.

One mistake the Windhoek hospital had made was keeping the bandages wrapped tightly around the burns, allowing them to fester and seep. Marieta had witnessed this disgusting process firsthand, watching the wounds ooze so much that they dripped onto the floor. At the South African hospital, they removed all the bandages, cleaned the wounds with sterile water, and wrapped them in a clear plastic. Then they lay the men on water beds to soften the chafing against the wounds.

The disadvantage in moving her loved ones to South Africa was that Marieta couldn't see them very often. It was too far to drive, and she was needed back on Harnas to manage the family's affairs—including the new farms inherited by Nick. Until she could arrange some time off, Marieta's friend assured her that her daughter, Jeanette, who lived near the hospital, would be happy to visit them and make sure they were getting the proper care.

Jeanette was just what the patients needed. Because Nick was so tightly wrapped and his hands were so badly burned, he couldn't even hold a book or magazine to read. Jeanette was a bit mischievous and brought some men's magazines when she visited. She would hold up pictures of naked women in front of Nick. Nick would smile and say "No! No!" but Jeanette persisted and soon had him laughing at her attempts to entertain him. She became a much needed distraction for both father and son, and she made the time go faster.

Finally, after two months of separation from her husband and son, Marieta arranged to go to South Africa. While they were definitely improved since the last time she had seen them and soon would be discharged, some disheartening news about Nico awaited her. His left hand was partially paralyzed. He had lost the tip of one finger and his hand shook nearly all the time. Both Nick and Nico had scars on their faces and limbs, but like the tough African men they were, they moved past it psychologically.

So little remained of the plane's contents that all the items fit in a small cardboard box that Marieta has kept. The fickle nature of fire shows up in the items that were saved. A Bible survived, although partially burned, melted, and blackened. Metal items fared the best: a silver engraved cigarette case, a tin coffee cup with still-visible pictures of zebras running around the edge, a matching zebra plate and utensils, a metal eyeglass case, a partially melted camera, pieces of Nick's handgun, and—a fire extinguisher. Other items survived that are less easy to explain: a half-melted bar of soap on a rope, a pack of cigarettes from the pilot's pocket, Nico's sock—although his other clothing was either burned off or melted onto his skin, and his canvas shoe with a melted rubber sole. This assortment of odd objects reminds Marieta of the inexplicable hazards in life and the inability to plan beyond the present.

After five trying months, father and son returned home to Harnas and resumed their lives. Nico returned to school, Nick picked up where he left off working on the farm, and as always the animals kept coming. The tragedy of the airplane crash couldn't stop the work of this family when there was always so much to do and so many animals that depended upon them.

Marieta and hyena Gumbi

chapter six

ONE ANIMAL *at a* TIME

The question is not, "Can they reason?" nor, "Can they talk?"
but rather, "Can they suffer?"

—Jeremy Bentham

AN ANECDOTE TELLS OF A WOMAN WHO COMES UPON the ocean shore and sees thousands of starfish, beached on the sand and slowly dying in the hot sun. Frantically, she starts picking them up and tossing them back into the sea. A man comes along and says, "Why are you bothering? There are thousands of them and you'll never save them all! What you're doing is futile. It simply doesn't matter in the larger scheme of things." In response, she throws another struggling starfish into the surf, points to it, and says, "It mattered to that one."

Some animals are cuddly, some are scaly, some are scary, and others are slimy, but Marieta believes that saving one animal at a time, regardless of how people feel about that species, is one of the most important tenets at Harnas. An example that proves the

rule occurred in 1999, when a group of nine brown hyenas—
each rescued separately from traps—came to live at Harnas.

Hyenas are often considered vermin that need to be exter-
minated. Of course, Marieta felt differently, believing that each
animal has a place in the cycle of life. Usually, brown hyenas live
solitary lives in the wild or join matriarchal clans. Harnas did
not have enough enclosures to provide one for each individual,
so they were living as a group in a large area near the food prep,
digging dens beneath the foundation of one of the buildings,
and breeding.

"One night," Marieta tells, "a friend of mine was out for a
walk and found a hyena pup out by itself, most likely rejected
by its mother, a member of the original group living under our
building. This friend brought him to me. He was just a tiny pup,
and I named him Gumbi. He lived for the first four years of his
life walking free in my garden, but then he started breaking into
the house—over and over. I knew he was getting to the age when
he might become dangerous, so I put him in his own enclosure
and visited him every day.

"Meanwhile, the other hyenas were still furiously digging dens
under the building. One night we all heard a huge *crash* and *bang*
and *boom* that went on for what seemed like minutes—like an
earthquake hitting! We ran out of the house and saw that it was
no earthquake. It was the floor above the hyena den collapsing.
So many of the hyenas died in the disaster! Such a tragedy! I
decided that this living situation must end and the remaining
hyenas must go out and find their own homes. I opened the
gate and out they went into the bush. At first they came back
each night, but eventually they adapted to their wild life and
disappeared. Since Gumbi was in his own enclosure, he wasn't

involved in the collapse, but I decided to try to do the same with him, driving him out into the bush. I took him far out in my truck and released him, but he ran after the truck when I left him. It broke my heart to see one of my babies so distressed, so I brought him back home."

Gumbi is another of the Harnas animals who has appeared in films and documentaries shot out in the bush. Despite opportunities to run away during filming, he never did. Apparently he had decided that the van der Merwes were his clan and he wanted to stay with them. Because so few of the animals raised on Harnas can be successfully released into the wild, Marieta knew the chances for Gumbi were slim, especially because he knew and trusted humans, so she accepted him back and created a permanent enclosure for him with hills of dirt to dig in and concrete-lined tunnels he could hide in when he wanted privacy, as well as trees for shade and his own water hole.

The Bushmen workers at Harnas usually feed Gumbi, because volunteers are not generally allowed inside his habitat, even though he seems relatively tame. He looks a bit like a sweet German shepherd from the front—about the same size with pointed ears and a long dark mane across his shoulders and back. From the side, however, his sloping back clearly marks Gumbi as a hyena.

In the wild, hyenas cause great fear because of their bite, one of the most powerful in the animal kingdom, with jaws capable of applying several tons of pressure. Hyenas are scavengers who actually do a great service, eating dead animals before they rot, thereby controlling the spread of disease. Their taste for food ranges more widely than most people believe. Ostrich eggs are a delicacy for them, but they'll eat most human food as well—and anything in between.

With such a reputation, Gumbi causes many people to give his enclosure a wide berth. One volunteer, Amy, however, learned a great lesson of compassion and friendship from her encounter with Gumbi. Animal Planet took a special look at their relationship for a short film they called simply "Amy and Gumbi." Whenever she had free time, Amy sat outside his enclosure. Gumbi always came over to the fence, lay down, and looked directly into her eyes. She admits he was her favorite animal and claims she could "sit with him forever and have full conversations with him that he seemed to understand." After Amy had established a relationship with him over a few weeks, Marieta allowed her to enter the enclosure and give him a dish of milk. While he was drinking, Amy petted him and ran her fingers through his thick coat, which she describes as feeling "like a horse's mane."

Successful experiences like this one can change a person's view of wildlife forever. One of the animals with the most dangerous bite, despised by many as vermin, depicted in children's stories— for example, *The Lion King*—as a conniving villain, becomes for this volunteer "beautiful" and "awesome." Who would have believed it—a young woman and a hyena, best friends?

Other animals that were first healed and then released have also chosen to return to Harnas for one reason or another. Doo Doo, a tiny white-faced Scops owl, now lives in Marieta's kitchen. When he was a fledgling, Doo Doo was found on the ground by some Bushmen who picked him up and brought him to Marieta because they knew the owl would not survive the night alone. Marieta raised Doo Doo to be released, but he refused to go. At less than six inches tall, he lives inside because the outdoor cats stalk him.

Once, he flew out of the house and landed in a baboon enclosure. The baboons almost killed him—put a hole in his head and tore his eye out—but a vet put him back together and saved his eye. His feathers grew back, and he lives on, sitting on the curtain rods or the ceiling fan blade most of the day, content to live with people. Instead of the expected "hoo hoo" that comes out of most owls, Doo Doo roars like a tiny lion when he is disturbed or hungry. He entertains himself by dive-bombing people and animals while they are eating, retrieving extra food for himself.

Another bird—but much larger—is named Asem, a white-backed vulture confiscated while being smuggled into Namibia. Asem came as one of 12 vultures brought illegally into the country in order to be killed for their body parts, which are used in Chinese medicinal remedies. By the time the birds were found, three were already dead from suffocation, but Marieta raised the other nine to be released. Asem, however, liked Harnas too much, and he kept returning. Unfortunately, on one of these return visits, like Doo Doo, he landed in the baboon enclosure. Baboons are not savage on purpose, but when a bird drops from the sky into their enclosure, they regard it as an opportunity for play. Though a large and powerful bird, Asem was no match for the troop of powerful adolescent baboons. They broke his wing, removed his feathers, and sorely damaged his dignity, but he survived and was found, removed, and taken to the vet. After that incident, Marieta decided to protect this bird that preferred living in her bird enclosure rather than flying free.

Asem rose to fame in the film *Beyond Borders* (2003), starring Angelina Jolie. Because he was so tame and comfortable with people, he was the perfect bird to use in the movie. Without

Asem, Jolie might never have heard of Harnas and become its patron. Since the movie was made, Jolie and her international family have occasionally visited the van der Merwes to relax and enjoy the animals. As the official patron of Harnas, she has generously donated over a million dollars to pay for electric fences that further strengthened the enclosures for the largest carnivores on Harnas. All thanks to a vulture.

Several vervet monkeys are housed near the large bird enclosure. Although this population changes as some become independent enough to join the troops of wild vervets that live by themselves in outer enclosures, some remain near Marieta's house so that she can keep a close eye on them for the rest of their lives. Vervets are small silver monkeys with black faces and teeth sharp enough to sever their enemies' arteries in an attack. The ones here in the garden, though, only bite the ears of volunteers gently, in play, to test them and see if they will scream.

Audrey, old and blind, is one of the monkeys that will never leave the garden area and join the wild vervets. Her story, like many of the animals at Harnas, is a sad one because it could have been avoided. Marieta tells it with mixed feelings.

"Audrey arrived here with her brother in 1988. A man phoned and said he didn't have a place anymore for his two pet vervet monkeys. I took them, of course, but when they arrived, I could tell there was something wrong. Eventually I realized they were both blind.

"I called the former owner, and he admitted that the monkeys grew up in a very dark room—probably a basement—and when he decided to make a bigger brighter room for them, he brought them out into the full sun all at once, and they went blind! Foolish! Foolish! *Och*! People are so bloody stupid! The male died shortly after arriving at Harnas, but Audrey has had a long life here. She lives in her little house, coming out to search through the pockets of visitors for peanuts. Sometimes she just likes to hold their hands."

Harnas contains more baboons and other primates than any other kind of animal, mainly because they outgrow their cuddly pet status. They look so much like us that we mistakenly believe they will act like us. Once a monkey is raised to be a part of the household, it becomes domesticated—sometimes picking up destructive habits from us human—and can rarely be returned to the wild.

Lonely soldiers found this out the hard way. In 1990, Namibia finally became free from the rule of South Africa, so the soldiers who had been fighting near the border of Angola, both against and with SWAPO, came home. While they had been living in barracks, far away from family and friends, many soldiers had taken wild animals as pets. The soldiers had tamed almost any kind of animal imaginable—vervet monkeys, bush babies, birds, snakes—but the majority of the pets were primates. But as the soldiers traveled south and reached the border of Namibia, they found they couldn't bring their pets home without the proper paperwork and a license to own a wild animal. The border police confiscated and shot the animals, which were tame and couldn't be released.

A friend of Nick's from his days on the antiterrorist squad called him in desperation and told him what was happening. Marieta and Nick jumped into one of their largest trucks, which held several different kinds of cages, and headed north to the Angola border crossing. When they arrived, they found animals still waiting to be shot, so the couple took them all, showing paperwork that proved they could own wild animals. They left for home with seven vervet monkeys, several bush babies, and lots of big birds.

Some of these animals had adjustment problems related to their having lived with soldiers. Every afternoon Adry, an old vervet monkey, would start screaming and acting crazy. No one knew what to do. Then one afternoon Marieta happened to have a glass of whiskey in her hand as she passed Adry's enclosure. The monkey reached out through the cage bars and screamed louder than Marieta had ever heard her scream. With infallible instincts, she handed the glass of whiskey to Adry, who drank it and immediately quieted down, content for the first time since she had arrived.

Marieta told Nick about the encounter that night, and Nick—who'd worked with soldiers during his time in the antiterrorist force—simply nodded.

"Soldiers and police—you know how they are," he explained. "In the evening they drink in the barracks, and they think it's fun to give their animals alcohol, too. So probably Adry drank every night with her owner. Now she's probably an *alcoholic* vervet monkey."

Some might argue that Marieta should have broken Adry of her addiction, despite her age and the pain she would suffer during withdrawal. Others might contend that Marieta should have put the monkey down. But she did neither. Adry was old, had no teeth, and probably wasn't going to live long. So Marieta did what she believed was the most humane. Every evening she gave Adry a shot of whiskey. The monkey was much happier, calmed down, and lived for another year.

People who come in contact with the blind Audrey and hear her story are often moved to sympathy and feel an urge to make a difference themselves. To feel for an animal, to sense its pain, and show compassion for it—and by extension all other creatures, both human and animal—forever changes them. That change is reinforced with every enclosure they look into—the "AIDS" lion, the Down syndrome baboon, the motherless meerkat, the three-legged wild dog, the orphaned hyena. The cycle of compassion between visitor, animal, worker, and volunteer is continuous, so wouldn't it be a shame if inspiring animals like these were put down just because they're not perfect? Marieta says, "If we had put down any of the animals we've found or been given, people would miss out on the understanding and compassion that their lives and stories awaken."

Like each starfish the woman throws back into the sea, the compassion of humans for these animals matters to Sam and Boertjie, to Audrey, Gumbi, and Tommy. At Harnas, each animal is perceived to have a soul—and each has a right to live. Through the handicaps they bear, each touches the lives of so many others.

Baby leopard Missy Jo and baby baboon Houdini play in Marieta's garden.

chapter seven

TO ERR IS HUMAN

*The least I can do is speak out for those who
cannot speak for themselves.*

—Jane Goodall

ONE TRUISM MARIETA REPEATS TO EVERYONE WHO
works, volunteers, or visits Harnas is that when an animal "mis-
behaves," the behavior can usually be traced to human error—
a gate left open, electricity turned off, and especially, incorrect
demeanor around the animal. People can have an inaccurate
perception of animals that are raised with people. Wild animals
are still wild by instinct—no matter how latent. Volunteers and
workers are given very specific instructions about how to react to
and treat animals. Every species is different and therefore requires
different training principles.

For example, a cheetah will respond to a forceful voice, a
direct stare, and an aggressive stance. He will almost always back
down if he feels you are more dominant. A baboon, on the other

hand, won't put up with such nonsense. If you are threatened by a baboon, you shouldn't meet his eyes. Look down—and if he still seems about to attack, you should look and point at another baboon. That will draw his attention away from you and he will most likely leave you alone. If you're holding something he wants, toss it away from you, and he'll forget about you long enough for you to get away. He probably just feels possessive and curious, not aggressive.

Lions are easy: they let you know immediately how they feel, and you should watch them carefully and react accordingly. Marieta claims she is not afraid of lions. "I believe in them," she says. "They will first warn you. Their whole face looks different when they're angry at you." When facing off with a lion, Marieta explains, "You should look him in the eyes, talk quietly to him, remind him you are a human, not an antelope or some other tasty treat, and then you should back away very carefully and slowly. A lion would prefer *not* to attack you, and if given half a chance, he'll leave you alone."

A leopard—well, a leopard is a cat of a different spot. There's a reason why, of the big cats, leopards are the most successful hunters. If a leopard truly wants to attack you—which he probably won't since you're not as yummy as his favorite meal, the impala—you won't see him coming. And if you do see him before he attacks you, you can't do much except admire this beautiful and stealthy predator in the split second before you say a final good-bye to the world.

No matter which animal is involved, several rules are universally sacred around Harnas. First, don't scream—unless the situation is dire, life and death, even. If the staff hears a scream, they take it seriously, and everyone will come rushing to your rescue. Second, don't run. If an animal sees you screaming and running, you are acting like prey, and then the animal will give chase—either in practice or for real. Most people have seen this primal reaction in domestic animals. Everyone knows that if you play with a dog by running away from it, it will chase you. If you drag a toy along the carpet, a kitten will pounce. A dog and a cat can get along famously—unless the cat runs, and then the dog simply can't help itself, and the chase is on. The instincts for predator and prey are stronger than any habits humans have instilled in an animal. All the romantic ideas about "tame" wild animals won't save you from that simple fact.

At Harnas, this natural reaction of a predator to give chase can have much graver consequences than with a domestic pet. If a volunteer is out walking with a lion, she might get scared if the lion turns and appears to confront her, and so the volunteer follows her first and most natural reaction: run. The lion will chase and pounce, and 500 pounds of lion can do a whole lot of damage, certainly more than a 50-pound dog that wrestles with you on your living room floor. Even if the lion is just playing, he can scratch, play-bite, and knock you down. Our fragile bodies and paper-thin skin are very unlike the thick pelt of a sibling lion he would naturally wrestle. The animal doesn't comprehend that his ordinary play can cause critical damage.

Many a volunteer has gone home with scars—usually not too serious, and sometimes worn with pride—but a more severe injury is always possible. Marieta, Frikkie, Schalk, and everyone else in the family repeat these two rules over and over, and yet sometimes the instinct to run is too great. At these times Marieta muses ruefully, "It's always the animal who is blamed."

Sometimes following the rules doesn't prevent an attack, and in this case, a person must use her own survival instincts. Marieta found this out herself in 1991 when she was making an ordinary visit to a female warthog she had known since the animal was an orphaned baby. The warthog has been given the well-earned title of Ugliest Animal in the World because of its enormous, out-of-proportion head, stiff whiskers, and fierce tusks. It looks like a creature that might have shown up in the bar scene in *Star Wars*. But Marieta had raised this female warthog, Porkie, since she was a piglet, and like most mothers, Marieta thought her baby was cute. Unfortunately, Marieta entered Porkie's enclosure not knowing one important fact: Porkie had gone into heat. So when Marieta knelt down where the warthog was resting, scratched her under her chin, and asked innocently, "How is my little Porkie today?" she had no way of knowing she was setting herself up as competition for the male warthog across the pen.

The male in the enclosure was in a position of protection over his fertile partner, a male that stood two and a half feet tall at his shoulders and weighed over 200 pounds. Without warning, he attacked from behind, driving his tusks into Marieta's buttocks and throwing her into the air. She landed with a painful thud, but before she could even process what was happening, he hit her again, sending her flying. Before she could get up or turn around, the male attacked again and again, jabbing his tusks into her and throwing her before she had any chance to put herself in a defensive position. Marieta felt the tusks enter her flesh each time, but she was helpless to do anything to help herself. Then there was a brief pause, long enough that she could turn to face him. In her delirium, she believed that her only chance was to catch him by his tusks and wrestle him down, but when he came at her again and she tried, he was too fast, and she missed.

"Then he became angrier because I was fighting back," she says, showing the scars on her arms and legs. "He attacked my face, arms, breasts, and legs now that I was facing him. I was bleeding everywhere, and I had only one thought over and over in my head. 'He is going to kill me. This is how my life will end.' He charged me one time, catching my calf and severing a muscle, making it impossible for me to stand up. I screamed for help, but the warthog pen was far away from our house, near the road, and no one could hear me. So I did the only thing I could think of. I rolled myself into a ball and pretended to be dead.

"He ran at me a few more times, jabbing his tusks into my back and behind again, and then he stopped. I guess I convinced him he'd finished me off because he began the ritual of marking his kill. He kicked sand over my body, covering me. I never lost consciousness and knew exactly what was happening to me. I felt every wound bleeding out, but I stayed as still as I could. My life depended on two things: the warthog believing I was dead and so leaving me alone, and someone rescuing me before I bled to death."

Out on the road, a water truck arrived to get its regular supply of water from the dam on Harnas property. The driver saw something red—Marieta's bloody shirt, no doubt—and a warthog kicking sand over that red object. Intrigued and worried, he drove to the house to report what he had seen. Simultaneously, Bushmen workers saw Marieta and began screaming. Sixteen-year-old Schalk was alone in the house, heard the uproar, and charged out the door, not knowing what was happening, but running nonetheless toward the sound. He found one of the Bushmen throwing rocks at the warthog to force him to the far corner of the pen. Schalk joined him and managed to drive the animal far enough away that he could dash into the pen, scoop up his bleeding mother in his arms, and rush her to the house.

He set her on the table and radioed Nick, who was working far out on the farm. While he waited for his father, he tried to clean and bandage Marieta's wounds. By now she was drifting in and out of consciousness, and while she was awake, she told Schalk, "Go back to the pen and make sure the male warthog doesn't hurt anyone else. Please don't kill him, though. It was my fault, not his."

Schalk ran back and found the warthog still snorting and kicking his legs out behind him, charging anyone who got near. Schalk made a decision to put the animal down—not because of what he'd done to his mother, but because the warthog was a danger to other people. The decision wasn't easy. It's difficult to sacrifice one animal to save others, but it's the kind of decision that sometimes has to be made in a wildlife sanctuary. Knowing the killing would make his mother sad, he nonetheless shot the warthog with his rifle.

When Schalk returned to the house, he found his father had arrived. They could see that Marieta's wounds were worse than they had imagined—and certainly more than they could handle medically at the house. By then the family had their own small airplane, so they put Marieta in it and flew to Gobabis. They radioed ahead and an ambulance picked them up at the airstrip and drove them to the hospital.

"I remember arriving at the hospital," Marieta says, "but then I blessedly lost consciousness."

Doctors worked on her feverishly, giving her blood, stitching up her many punctures and tears. Her teeth were broken out on the left side of her face, so the doctors put them back in with wires and pins. Some wounds, though, were beyond repair. The warthog had jabbed her left ear, making her nearly deaf on that side, and he cut the nerves on the left side of her face, leaving her with no feeling there.

"The attack left me with many scars on my body but only a few on my face." She points to a small line above her lip and another on her eyebrow. "I protected my face fairly well by rolling into a ball. The severed muscle in my leg hurt me for a long time after I left the hospital, and continued to swell, eventually requiring a temporary plastic pipe to be inserted into my leg to drain the fluid. When I finally regained consciousness, I remembered nothing about the attack and I didn't even recognize my own children or husband! I stayed in the hospital for three weeks, and at first all I could remember was the nightmare of the warthog bumping against my head and body. Each time I closed my eyes, he was there again—*bump bump bump*—slicing tusks into my skin."

Slowly Marieta's memory returned, she recognized her family, and she remembered the details of the attack. Schalk was also showing signs of shock, having been the one to rescue his bleeding mother and kill the warthog. In fact, the whole family was overwhelmed by this jarring attack. Amazingly, Marieta suffered no damage to her internal organs, despite being hit numerous times from behind. When she finally improved enough to laugh again, the doctor told her she was lucky she didn't have a "skinny bum" or the tusks would have pierced her kidneys. Now when she tells the story, she brags that her "big bum full of holes" saved her from further damage.

Does Marieta hate warthogs now? After the attack, did she go back and curse and kill the warthogs she still had? Of course not. Every time she tells this story, she is careful to note, "It wasn't the animal's fault. I went into the enclosure when I was not supposed to." Human error—although certainly not a conscious error—led to the attack, nothing more. The male warthog was following his instincts, protecting what was his. Marieta was just the unfortunate victim of an age-old natural cycle.

As evidence of her forgiveness, these days a large warthog named Murray wanders the area where guests stay, trailed by his harem of three smaller wives. He will allow people to pet him—especially if they offer him a piece of fruit—and at night that snorting you hear outside your window is most likely Murray, checking to see if any of that fruit is available. Some people are nervous when they see him—especially if they know Marieta's story—but Murray would rather avoid a confrontation. When the females are in heat, he takes them far away.

Some of the injuries are inflicted on other animals rather than humans. In 1993 a litter of four lions was born that would have a big impact on Harnas. They were the offspring of Elsa and Schabu, and Marieta named them Sher Khan, Savanna, Teri, and Esria—all names of animals and people significant in some way. Sher Khan, for example, is the name of the oldest lion in Etosha National Park, and Savanna is the name of one of the first donors to Harnas.

Elsa was still very young to be a lion mother, so the cubs were taken and hand-raised. Marieta always tries to hand-raise the lions on Harnas because they are so much easier to handle once they grow up, and if they need veterinary care, that familiarity and trust in humans pays off. The four cubs grew up in the courtyard area right outside Marieta's kitchen. Sher Khan was the only male, aggressive and dominant even as a cub—not the case for many of the male lions raised on Harnas. At feeding time one day, as Savanna ate, Sher Kahn decided he wanted her portion. In the squabble that ensued, Savanna fought back, and Sher Khan swiped at her with his claws extended, slashing out one of her eyes.

Savanna was rushed to the vet in Windhoek, where the better vets had their practices in those days, but the vet could do nothing for the eye, so he sewed the space closed. Savanna recovered and learned to compensate for having only one eye. She was so young when she lost it that she made an adjustment that would not have been possible had she been any older. She was a gentle lion, easy to handle, loving to her human family. She was finally placed in a large enclosure and lived in peace for many years. Then she got into another fight with Sher Khan and ended up with a wound in her leg that eventually, despite much medical intervention, developed a yeast infection that killed her. She was only 11 years old, relatively young for a lion in captivity to die, and her gentle soul has been missed on Harnas.

The 1990s was a tough decade for carnivores, especially lions, on Harnas. Marieta recalls, "In 1997, a neighbor offered us the carcass of a cow that had run into a fence and broken her neck. I accepted gratefully—especially pleased because we were short of meat at the time. What I didn't know and what the neighbor didn't understand is that the cow had been recently inoculated for anthrax, an intestinal disease that affects livestock. Two days after feeding the beef to the carnivores, animals on Harnas started dying. Lions, leopards, cheetahs! I was frantic, couldn't see any reason for the deaths, and so called the vet. He came to Harnas, dissected the dead animals, and diagnosed the problem as anthrax poisoning. Then he gave injections to the animals that were still alive but had eaten the meat."

All of the remaining animals survived, but Marieta had lost seven of her babies. One of the animals who initially survived was a magnificent lion called Hemingway, named after the author Ernest Hemingway, who was well known for his adventures all over the world, including Africa. The lion Hemingway was a cub that had grown up in Marieta's house and slept in her bed. He had developed into a massive lion that lost his cute, cuddly ways, and became an awesome sight for tourists as he rushed the fence and roared in dominance. Marieta did everything she could to save this marvelous creature.

Although Hemingway survived the first bout of illness, the anthrax had attacked his liver and kidneys. Eventually he became so ill that the vet suggested they put him down.

"No," Marieta decided. "If he is alive, we must give him a chance to beat this."

And he did beat it. The vet had to remove a kidney and a half, but Hemingway lived. His diet and general health had to be watched very carefully—although looking at him, you'd never guess he had any health problems.

"But Hemingway got sick again a few years later," Marieta says, her voice full of the memory. "A cow's udder was given to him as part of his diet but it had too much fat in it. As a matter of fact, it had too much fat for many of the animals, and their teeth turned a sort of yellowish color, a symptom the vet used for diagnosis. The vet told us that the animals could have no more udders, and Hemingway must be given only the cleanest, leanest red meat."

All the other animals got better quickly, but the remaining half kidney could not clean out Hemingway's system. When he died, the family grieved as if he were one of their children.

They would recover and move on, however, for life and death are closely related in the captive care of wild animals, as they are in the wild.

In the feeding of carnivores, one animal dies so that another can live. In North America and most other industrialized countries, people buy their meat in packages in the supermarket, and so the life and death of the animal offering that food is rarely considered. But in Africa, people learn early that animals die and others are born as part of a necessary cycle—and that cycle is completed at a much faster pace than in the human population. Most of the time people who work with animals understand this, and even though they grieve, they comprehend the necessity and inevitability of death. But sometimes death at Harnas is more painful, and the family grieves more intensely. This was the case in a lion tragedy that struck in the 1990s, and once again, it was the result of human error.

In February 1995 a litter of lion cubs was born on Harnas, and their mother deserted them for unknown reasons. So Nick and his daughter, Marlice, became the foster parents of the cubs. The runt of the litter was christened Mufasa, meaning "king" in Swahili. He had huge feet and ears and seemed unsure of how to mobilize his out-of-proportion limbs. The family fell instantly in love with the clumsy cub, especially Marlice, who claimed him as her own.

Mufasa eventually grew into his ears and paws, growing larger than all of his siblings. He had the run of the house, followed Marlice everywhere she'd let him go, and slept in her bed with her. As he grew, other people—specifically, the workers at Harnas—became wary of Mufasa. Nick found out later that they often kicked him, and so the lion became aggressive when he saw those who bullied him. He remembered every kick and punch—and who had dealt them.

With Marlice, though, he was as gentle as could be. He kept his claws retracted at all times when they played, and if he misbehaved, she disciplined him gently. Mufasa responded to her as if he were still a 10-pound cub rather than the 500-pound hunter he had become. Visitors were amazed seeing a beautiful young blond girl walking alongside a huge male lion, scratching him under his chin and even occasionally riding him. Marlice, a girl who lived isolated from other people, considered Mufasa her best friend.

This peace and contentment changed in April 1997 when Mufasa was just over two years old. Marlice took the motorbike to attend to a problem quite a distance from the main house and Mufasa's enclosure. Marieta was doing some gardening and asked a Bushman named Abraham who was assisting her—along with nine Bushmen children—to take some branches she had cut to the fire pit for burning. Rather than taking the normal but longer route around Mufasa's enclosure, Abraham took a shortcut through the lion's territory. The nine children followed him through the gate and watched him latch it.

The group began walking across the open space in the middle of the enclosure, unaware that Mufasa was lounging quietly in the shade of his wooden play platform. Suddenly Abraham saw Mufasa, screamed, dropped the branches, ran back through the gate, latching it from the outside. He left the nine children to face the lion alone. Once the children realized what had happened and saw the massive lion observing them with interest, they did what was instinctive but absolutely wrong. They began screaming and running in all directions.

Mufasa, normally passive and sweet—Nick called him the "most gentle of all the tame cats on Harnas"—reacted aggressively, forgetting his human training and remembering his abuse at the hands of the Bushmen. His system began to work purely on instinct. He became a wild lion with a total of nine "prey" animals in his sights. When something runs, chase it and bring it down, his brain told him. He began the chase and the children scattered, screaming.

By this time Marieta had heard the screams. Dropping her gardening tools, she ran to the enclosure, entered quickly, and put herself between the aroused lion and the hysterical children.

"When Mufasa pounced," Marieta begins the story, "his jaws closed around my left arm, but I knew I had to get those children to safety. So I used my other arm to pick up children one by one and toss them over the tall fence. Adrenaline helped by masking my own pain and giving me strength. And it helped that Bushman children are very small and light. I tossed the eighth child over safely, but before I could get the last little boy, Mufasa let go of my arm, which was shredded and bleeding, and snatched the boy around the waist. Then Mufasa paused and looked right at me. It was as though he was challenging me to stop him."

At this point Marieta made a final effort to save the situation. "I touched Mufasa's enormous head and began to talk to him gently, pleading with him to let go of the boy. I thought I could calm Mufasa and remind him of his gentle upbringing in our house where everyone loved him. Mufasa looked at me and seemed to be listening, but he refused to release the boy. He couldn't find his way back home."

Nick, having heard the commotion, arrived with his rifle. He raised it, sighted Mufasa's forehead, and was about to pull the trigger. He froze. Mufasa was his favorite lion and his daughter's best friend. He couldn't bring himself to do it. Marieta continued to talk to Mufasa—who was doing no further damage to the boy, just holding him firmly—but the only response he gave to Marieta was a growl. She realized, hearing this sound, that Mufasa was too much in the world of wild lions, too far gone to pull back. She pictured her life's work being destroyed if this little boy was killed by one of her lions. She could see the negative publicity, the public outcry against keeping wild animals contained, and the subsequent government edict that might close Harnas and take all her babies away from her.

She looked once more at this magnificent lion that had played in her kitchen and cuddled with her daughter. Then she looked into Nick's eyes and gave the sad command: "Shoot him."

Nick fired a first shot and then once again, for Mufasa was not a small animal. The top of the lion's head blew off, the bleeding boy was released, and Marieta tried to staunch his bleeding and her own while Nick called for the Med-Rescue aircraft in Windhoek.

Far away, on an outer edge of the farm, Marlice heard one shot and then, shortly after, another. With the second sight that mothers claim to have in relation to their children, she felt her heart squeeze and drop. She jumped on her motorbike and fled home. Dropping the bike outside Mufasa's enclosure, she ran in and threw herself across his body, drenched in blood, just in time to feel his last heartbeats.

Her heartbreak led her to shout at her parents, "How could you do this? I hate you!" Unable to answer their daughter's questions with any response that would satisfy her, knowing her pain far surpassed their own, but understanding the necessity of killing this one lion to save many more—as well as saving the boy's life—they remained silent.

When the aircraft finally arrived, the injured boy was loaded on, but Marieta—realizing the need to spend time with her daughter—elected to go by car to Gobabis. Nico was away in Pretoria and Schalk was with the team, playing rugby, so she knew that Marlice would have to drive her. Not much was said on that drive, but Marlice saw her mother's bleeding arm, ripped by the lion as Marieta had tried her best to save everyone involved. After the doctor sewed up Marieta's wounds with numerous stitches, both mother and daughter were sedated. It wasn't hard to see that Marlice was in as much pain as Marieta.

The Namibian press seized on the story, as newspapers will, always looking for a gory tale to repeat, and countless letters were written supporting both sides—to contain wild animals or not. Finally, in hopes of ending the controversy, Nick wrote a letter to the paper, outlining the events of that day and making it clear that "the incident was not caused by Mufasa." He ended his letter with a defense of Harnas and of the conservation of Namibian wildlife.

"All the animals on Harnas would have been sent to their doom by a world that believes that the animal must continually make space for man. By a country that has an attitude towards nature that is deplorable. A nation whose values towards that which is the only good left on this earth makes me sick." He ended by imploring readers to consider living in harmony with wildlife rather than destroying it in pursuit of their own selfish desires. "Can we not live in peace with [the animals]? Or will we only be happy once we have it all and they are destroyed?"

Marieta takes a group of volunteers for a baboon walk

chapter eight

MONEY *and* CHANGE

*If all the beasts were gone, men would die from a great loneliness of spirit,
for whatever happens to the beasts also happens to the man.
All things are connected.*

—Chief Seattle

NICK AND MARIETA FOUND THAT FINANCING A GROW-
ing enterprise like their animal rescue program was an unending
challenge. One by one, they sold off their cattle farms to pay for
what was once called Marieta's hobby but what was becoming
the family's mission. Perhaps the hardest property to sell was
Tennessee, the farm where Marieta had grown up, but even
that was sold in 1998. Times were difficult for white Namibians who sold land after independence came in 1990 marking the end of apartheid in both Namibia and South Africa.
The 1995 Agricultural Land Reform Act stated that before a
farm could be sold, the government had to be informed and be
given the first option to buy the property. If the government
bought the farm, it was usually given to a black citizen with

the understanding that he would have 15 years before he would have to start paying for it.

In theory this is a good way to offer the underprivileged the opportunity to improve their lives. Sometimes it worked, but sometimes the new owner took advantage of the system by letting the farm go to ruin, thereby lessening its value and decreasing the price he would have to pay to the government later. The government lost a lot of money this way—but so did the original owners, like the van der Merwes, because the government would pay only rock-bottom prices. The farms that Marieta's and Nick's parents had worked so hard to build were sold for far less than they were worth, but they had to be sold, nonetheless, to sustain the never-ending stream of animals making their way to Harnas.

A lion, for instance, eats the equivalent of seven horses or mules per year. With 16 adult lions on the farm, that means 112 animals have to be bought or raised as food for the lions alone, and that runs around $8,000 a year. The African wild dogs require about $20,000 a year for food, and the baboons (85 in total—and growing in numbers) eat around $91,000 worth of fruit and other foods. The list goes on and on: 7 caracals, 30 cheetahs, a brown hyena, 8 leopards, 2 crocodiles, 18 dogs, 45 cats, many birds, a milk cow, 16 jackals, a giraffe, and numerous sheep, goats, and chickens, and on it goes, with each animal requiring daily sustenance.

The wealth of one of the ten richest families in Namibia was nearly depleted, and Nick and Marieta knew they had to find new resources for the animals and the future of their family. In 1997, they created the Harnas Trust Fund, which protected their personal money, and they created the Harnas Wildlife Foundation, a non-profit organization that allows Harnas to receive money through sponsors and donations. Donors give money for general use, such as GPS collars, vehicles, antipoaching efforts, and the Wild Dog Project, which is an ongoing research program. A donor can adopt a specific animal or give money for general care of animals as needed. Another option is to "adopt" a specific child of the Bushmen, paying for his or her food, clothes, toys, education, and medical care, as well as preservation and development of the Bushman culture. Other monies go for enclosures and upgrades like better water holes or improved electric fences.

By creating the foundation, the family conceded that they couldn't finance Harnas alone anymore. Not a week goes by without some new animal making its home there. These are the luckiest of the animals among the many in Namibia that are homeless. If Marieta's dream was to be realized, it was going to take more than the family to make it happen. So in the year 2000, the last of the cattle were sold. The van der Merwes officially became full-time conservationists, and Marieta's hobby became the family's livelihood.

Still, running the animal sanctuary continued to stretch the family budget to its limits. Marieta and Nick decided that they would need to involve the outside world in other ways as well. In 1993, they opened Harnas to guests—tourists—who wanted to experience close encounters with the wild animals the family had raised to be so tame. The enterprise was informal at first, and guests moved freely around the wild animals in the same way that the family did. This was a bit of a surprise to some guests who hadn't realized that the advertised close encounters were very close indeed.

One night a group of guests arrived after dark and were escorted to their rooms. Then they were invited into the center of activity on Harnas at that time—the family house—where a billiards table had been provided to entertain them. The guests started drinking, playing pool, and having a good time. They were unaware, though, that one of the leopards, Patcha, was wandering freely around the grounds. Drawn to the noise of the billiard balls clicking against each other and rolling around on the felt, she arrived—stealthily, as leopards do—and the guests were completely unaware of her presence or of the fact that she was watching with great interest the activity on the table. Without warning, Patcha jumped on the table and pounced on a rolling pool ball. The guests sobered up in a hurry, yelping, and pressing themselves against the wall. Marieta strode in, lifted Patcha off the table, and carted her away, leaving the guests to gasp and finally laugh. They would have a memorable story to tell their friends at home.

Sometimes the guests who were surprised were family friends who had joined the van der Merwes for a visit. One day Schalk invited two large, muscular rugby players from his team. They had had a lot to drink, were feeling uninhibited, and took their shirts off. Suddenly, one of them caught movement on the roof of the house, looked up, and saw two leopards watching their escapades. The man never took his eyes from them, hypnotized by their intense stare at the human activity below, but he nudged his friend and pointed up. Both men did what seems to be instinctual when people see free-roaming leopards: they plastered themselves against the stucco wall and tried to escape by slinking along the wall undetected. The eyes of the leopards followed their every movement.

The two men got scratches all over their backs from the rough stucco wall, so closely had they pressed themselves against it, and some of the scratches were even bleeding. Apparently no one had told them that leopards, much like housecats, like to be as high as possible to watch things. They are habitual tree climbers, but since there were no trees nearby, they were lounging on the roof, observing the humans below.

Watching the whole drama unfold, Schalk was as amused as the leopards were entertained. He loved to see his rugby teammates blanch with fear, since they usually acted so brave and invincible. Rugby players are arguably the toughest men on the planet, and anyone who has watched a rugby match would probably agree that these men are fearless during competition. But this doesn't always translate to regular life—especially when one of your teammates is a prankster by nature.

It became a game for Schalk when his teammates visited Harnas to see if he could tear down the rugby toughness and give them reason to scream. For one such party, the players arrived in several different cars or trucks. Schalk had warned them over and over that if they were approached by a lion, they should *never* turn their back, nor should they sit or fall down because that would make them look smaller and easier to catch.

Fritz arrived first. He got out of his truck and started walking toward the gate when suddenly he found himself surrounded by four lions. They were part of a tame litter that Schalk had left roaming the parking area on purpose. They were not fully grown, but still waist-high and powerful—with those glinting teeth. Fritz edged toward the chain-link gate and grabbed on with both hands, but one of the lions inserted himself between Fritz and the gate, so Fritz had to lean over him in order to hold on to the gate. Then the lion began to move outward away from the gate, but Fritz refused to let go, assuming the lion would attack if he fell. As the lion backed away, he lifted Fritz right off the ground until he was, in effect, riding the lion. He froze there, hanging in the air above the cat.

In the house, a Bushman told Schalk that one of his friends had arrived. By the time Schalk sauntered over—a big smile on his face and two cold beers in his hands—Fritz had turned white with fear, and his hands were bleeding from holding on to the wire so fiercely. Schalk says Fritz was "swearing swearing swearing, and swearing some more," as only a rugby player can.

Schalk wasn't the only one who loved to turn rugby men into shrieking, cowering boys. Marlice, too, got in on the action. When Braum—a professional rugby-player friend—was visiting, he kept saying that he wanted to sleep out under the stars in the bush. Finally the arrangements were made, but—unknown to Braum— with a little added adventure, compliments of Schalk, Marlice, friend Robby, and Robby's brother. Before the campout, Robby had prepared for the prank by tearing up an old T-shirt and dousing it with animal blood from the food-prep area. He hid the shirt in the bushes near the site where they had chosen to camp. The site was a beautiful one—right by a water hole with a nice fire pit. They built a fire, grilled some dinner, and were lying in their sleeping bags enjoying the stars. Marlice and Robby's brother were hiding in the bush, videotaping the scene.

At an assigned time, Robby got up and said he was going into the bush to take care of some business. A few minutes later, Braum heard screaming and shot to his feet, still enclosed in his sleeping bag.

"Aaaaaaggghh! Help me! Oh, nooooo! Heeeellllp!" was what Braum heard. Then silence.

The expectation of the pranksters was that Braum would jump out of his sleeping bag, run toward Robby's voice, and find only a bloody T-shirt. But they had overestimated Braum's courage. At the moment the cries stopped, a wild horse, standing in the water hole behind Braum, snorted. He jumped, unable to place the sound and sure that whatever had attacked Robby was coming to get him next. He sprang out of his sleeping bag, grabbed the biggest burning limb from the fire that he could carry, and started sprinting in the direction of the house—about a 30-minute run if you don't stop to rest.

The videotape showing all this is almost too jumpy to watch. Marlice and her co-conspirator in the bushes were laughing so hard they couldn't keep the camera still. Once Braum hightailed it out of there, the rest of them had time to regroup, get in the truck, and drive around a different way so that when Braum arrived in Marieta's kitchen—breathless and terrified—everyone else, including the "victim," Robby, was sitting around the kitchen table, smiling calmly. Braum realized he had been royally duped. "Bastards!" was all he said before the laughter broke out.

Sometimes the tourists turn out to be predators, though, and the family found they had to watch visitors almost as closely as the animals. One story has a second chapter in Windhoek when a monument was built for the freedom fighters of SWAPO. Chinese workers were brought in by the hundreds to help complete the construction, and most of them stayed in a block of apartments in the city. When their work was finished, they left en masse. The apartments remained locked and supposedly empty, but one enterprising robber thought perhaps some Chinese had left valuables behind, so he broke into the building and began a search. Suddenly, he heard mewing—lots of mewing.

This was a robber with a heart, and he called the authorities to report that there was an entire population of cats left in the building. When the doors were all opened, they found hundreds of cats—grown cats, kittens, and everything in between. Some were dead and some were dying from starvation and dehydration, kept by the Chinese not as pets, but rather as food. This was no urban legend: some Chinese eat cats, and when they left the country, they thought so little of their "livestock" that they locked them inside their apartments—without food or water, never contacting the authorities or even releasing the cats to fend for themselves. Without the robber, hundreds of cats would have died instead of dozens. A call went out on the news to save the cats, and the good people of Namibia took them in.

This menu anomaly wasn't news to Marieta and her family. A few years earlier, a Chinese official had visited Harnas with his entourage and had a run-in with Schalk and Jo over a cat. Wandering the grounds are about 45 cats, at least two of which are African wild cats that have bred with the domestic breed. Sterilizing all of them has been a big project, as females in heat will find partners from the surrounding free-range population of wild cats and black-footed cats. (Although both species are wild, they can breed with domestic cats.) By October 2006, all the cats had been neutered, and the population was finally stable, other than rescue cats that occasionally are added to the group.

Although they walk freely, these 45 Harnas cats are not strays. All of them have names and established territories. They are fed at five different feeding stations by volunteers, and they receive medical care if needed. During cold nights, many tourists take the cats into their rooms to keep warm—and some do it just because they love them and miss their own pets far away at home.

Perhaps the queen of the cats is a long-haired, brown-striped beauty with golden eyes, Lady Godiva. She carries herself with nobility. She openly approaches anyone at Harnas, assuming friendship, and purring patiently while she waits to be petted and appreciated for her beauty and stature. So when Schalk and Jo found that the Chinese official had locked the tame Lady Godiva in his black Mercedes, windows up, they became very upset. Schalk was preparing to break the car window to rescue the cat when the official arrived back at his car.

"What are you doing with Lady Godiva?" Schalk asked, amazed at the brazenness of the man's behavior.

"I'm taking it," he said.

"But it's *our* cat. You can't just take it!"

"No," the official answered with the authority he was used to wielding in all situations. "This cat was roaming free. It is mine for the taking."

Schalk's temper rose, and he ordered the man to give him the cat, threatening him. "Take that cat out and give her to me or I'll have the police stop you on the road."

Whether because of fear of another Chinese "cat scandal" or Schalk's imposing physical dominance, the visiting official freed Lady Godiva from his car, dropped her on the ground, got into his car, and roared off. His entourage went with him, all acting offended—although it could have been a mask for embarrassment. Unaware that her life had just been spared, Lady Godiva padded away, looking for the next person to adore her.

Lady Godiva continued to live a pampered existence among those who loved her until September 2009, when one morning she simply didn't wake up. A good life on Harnas is the best she could have hoped for, and her regal beauty and sweetness will be remembered fondly by many.

Such a close-knit family didn't find opening their home and sharing their space and their animals always easy, even though it was financially necessary. Nick especially felt the loss of his privacy, and he wasn't happy with more and more strangers wandering his farm, but true to his nature, he remained stoic about the changes. He loved his wife, and he knew how much this endeavor meant to her. So the family persisted, and Harnas continued to grow and evolve. And the adventures were just beginning.

Nico and Melanie on their wedding day

chapter nine

PEOPLE PEOPLE

*Our task must be to free ourselves from this prison by widening
our circles of compassion to embrace all living creatures
and the whole of nature in its beauty.*

—Albert Einstein

WHILE NICK AND MARIETA CONTINUED TO CONVERT
Harnas for the purposes of conservation, the three van der
Merwe children were finding their own way in the world, meeting people who had never slept with cheetahs or wrestled with
lions. The world outside of Harnas was full of "people" people
rather than "animal" people, and the three children were learning
fast that their upbringing was unusual, to say the least.

When Jo Swanepoel married Schalk and came to Harnas to
live, she had never owned any animal except a dog, didn't eat
meat, and had grown up living safely in the city, the daughter of a minister. She wanted to help out, though, so on her
first morning at Harnas, she asked Marieta, "Can I help?"
Marieta's typically pragmatic response—not realizing her new

daughter-in-law's innocence in the ways of farm life—was "Yes, you can sort these intestines."

Jo admits, "The shock must have stayed on my face for quite a while."

Schalk and Jo met in Windhoek through a mutual friend, one of Schalk's rugby teammates. While Schalk was quiet and shy, Jo was gregarious and friendly. Whereas Schalk would stand in the corner at parties, sipping his one beer for the night, Jo would be dancing and laughing in the center of the action. But when Jo finally walked up to silent but handsome Schalk at one of those parties and introduced herself, both of them were smitten, and one year later to the day, they were married.

Schalk didn't bring girls home very often, so the first time he brought Jo to visit, Marieta knew it was significant. The two had been dating for about three months, and Jo had heard about the tight-knit van der Merwe family, and she was feeling nervous about meeting them for the first time. They arrived late on a Friday night and Schalk parked his truck, got out on his side, and started walking toward the bridge that led to the house. Jo gathered up her bags and got out after him. As she gazed out into the endless darkness, she distinguished eight bright-yellow eyes moving toward her.

"I turned around and four—FOUR!—lions were walking toward me. I'd never even seen a big dog, let alone a lion. I yelled 'Schalk! Schalk!!! SCHALK!!!' I thought, 'I'm not going to make it through my first night here!'"

The big cats, curious about this petite young woman, moved around her at chest level, rubbing up against her and pushing their gigantic heads into her and knocking her around—just testing her to see if she'd scream and run. Without Schalk nearby, she probably would have. Even so, she looked at him, panicked and pale, waiting to be rescued. Schalk smiled and said, "That's just Macho, Kublai, Ushiwayo, and Larato. They're fine. Don't worry about them. They're not even full grown." He took her arm and led her away from the moving walls of musty, yellow pelts smelling of grass, sweat, dirt, and fresh meat. Perhaps Schalk was also testing Jo to see if she'd scream and run.

Crossing the grass to the family house, Jo saw many more surprises: an ostrich or two, a baby springbok that startled her by performing its high-flying kick, and a small army of dogs that escorted them, barking and nipping at their heels. Schalk pointed his flashlight on the grass ahead of them to keep them from stepping on any of the tortoises that wandered freely, numbers painted on their backs to keep track of them so they won't end up as dinner on a Bushman's table—for whom they are a delicacy.

Marieta knew that any woman who married her son would need to be tough, so she didn't do anything special for Jo's visit. Normal chaos reigned throughout, especially in the kitchen. When Jo walked in, she spotted lots of animals and heard a lot of noise. Baboons and dogs competed with the audio from the television. In the middle of dinner, a baseball-sized bird swooped down and across the surface of the table. Jo flinched, but no one else reacted. Baboons ate off people's plates, and everyone talked, cawed, barked, and mewed at once.

Jo, raised in a quiet home, was overwhelmed, but she also felt she belonged. She thinks now she was a necessary addition to this chaotic household. While the van der Merwes are a passionate family who will argue, fight, and then easily make up, Jo tries to be a buffer, the calming influence. In a family of people who prefer dealing with animals, Jo works well on the telephone with potential guests, travel agents, and suppliers. She knows how to make people feel at home—even when that home is filled with lions and baboons.

Jo had to pass one other rite of initiation before she and Schalk could marry. Jo's parents had to come to Harnas and meet their daughter's future in-laws. Schalk, who had grown up thinking the best thing in the world was being out in the wild surrounded by huge carnivores, planned an outing for Jo's city-dwelling parents. He went out by a large gravel pit and prepared a campground for a night of sleeping under the African sky. He took out sleeping bags and cots, made a fire, and provided a big pot of stew. And as many men do when they go camping, Schalk brought along his favorite pet. His pet was not a loyal dog, however. Schalk brought along Savanna, the one-eyed lion, now full-grown (and pregnant) and very jealous of Schalk's attention to Jo. Savanna had come to think of Schalk as hers, just as many dogs feel about their humans, but most dogs don't weight more than 350 pounds and have three-inch claws.

Schalk realized that Savanna might end up wanting to be the center of attention, so to keep her busy while the four people enjoyed a nice dinner around the fire, he brought along the carcass of a sheep. Savanna snatched the sheep by the neck with her powerful jaws, shook it once to make sure it was dead, and disappeared into the bush. After some initial surprise at having a lion companion on their trip, and a dead sheep to boot, Jo's parents adapted to the adventure, and the night began pleasantly enough. Jo admits that, hard as it is to believe, they "sort of forgot about the lion. We were eating and talking and hadn't seen her, so we just forgot about her." When she was tired enough for sleep, Jo announced she was going to bed. Schalk had placed Jo's and his cots and sleeping bags down in the gravel pit, and Jo made her way down the slope to her cot, holding her flashlight steadily in front of her.

As Jo knelt down to shake out her blanket, without warning Savanna leaped from behind and jumped on her back. Jo screamed as she was flattened under the heaving lion, that began to nip and bite her—not enough to draw blood, just enough to send a message that Schalk belonged to Savanna. Schalk heard the muffled scream and came careening down the gravel descent. He tried to pull Savanna off Jo, but the jealous lion wouldn't let go easily. When finally Schalk convinced Savanna to move, the lion wandered off into the night, leaving behind the smell of sheep carcass from her heated breath. Jo settled down, distraught but determined to make the best of the night, despite the attack.

Just as she was drifting off, Savanna pounced again, and Jo felt herself crunched under a beast more than three times Jo's weight. Schalk was ready for her this time, recognizing Savanna's need to show dominance over Jo, and he was able to body-block Savanna partially. Off the lion went again, swishing her tail with defiance and haughtiness. Savanna hadn't given up ambushing, though. Jo says, "I was in tears, and all throughout the night, she kept jumping on me just when I thought I was safe." And where were Jo's parents? Sleeping in the car.

In time, Jo adjusted to life on Harnas and gave birth to two children—Samar, a boy born in 2002, and a girl, Aviel, born in 2005. Jo is a gentler influence on her children than Marieta or Marieta's father were on theirs. She tries to protect her children from "seeing death so close on Harnas." When asked if she would be as tough as Marieta and make Samar get back on a horse if he had fallen off, Jo, Schalk, and Marieta all answered in unison with wry smiles, "No!"

Schalk, on the other hand, believes danger lurks everywhere—in the bush and in the city, too. He argues, "If you really want to avoid all the dangers, you have to sit in a small room with no windows and close the door to the world. And if you're afraid, your kids will be afraid. I tell [my children] what to do and what not to do, but there's nothing more you can do without making them crazy. . . . I personally think you have to see the reality of life." Schalk has introduced life and death on Harnas to his son, and he tells him the truth about animals that need to be slaughtered to feed other animals. He states quietly, "I think it's better that they learn the truth instead of finding out the wrong way—or in a way you don't want them to see it."

Although Jo still prefers a day of playing with her children or working in the school, she occasionally sits on the lawn and strokes a cheetah or bottle-feeds a baby baboon. She can lead the visitors' tour of the adult animals in the outer enclosures and seems proud that she has been "bitten by every animal on Harnas." At first she felt useless at the sanctuary because all the skills she'd developed in her life didn't apply in the world of wildlife conservation, but eventually she came to understand what her role might be.

"God uses me to be the 'people person' in a family of 'animal persons.' It's interesting because all three spouses of the Harnas children—me, Rudie, and Melanie—are more people-persons, almost as if the family needed that balance. That influence has brought a lot of change in Harnas and the family."

Nico had also grown into a handsome, intelligent young man, dark where his brother was blond, but with the same well-built stature, confidence, and charming smile. Nico decided he wanted to go to medical school despite the serious and permanent damage to his left hand resulting from the plane crash. After three years of study, he finally had to admit he couldn't continue because his hand shook too much. His advisers encouraged him to try veterinary school because the precision needed in treating animals was not as exact. He moved to the vet school in Pretoria, South Africa, and was able to skip the first two years of training because of his medical school background.

With only one year to go, though, Nico became the victim of a mugging and was knifed in the shoulder. He healed completely, but he decided he wanted to be among his family for a while, so he took a year off and came home to Harnas. He was fed up with school—he had been going for eight years straight—and he just wanted some time to consider his future. During that year, he made the decision not to return to finish veterinary school. He realized he wanted to do something completely different. He loved music, people, and good food, so he moved to Windhoek and opened a restaurant. The establishment, which served traditional African food, did well, and Nico felt the success of doing something completely on his own. His employees and co-workers claim he showed signs of being a great leader, a man people would love to work for, people who would be charmed by his gentle management and loyalty.

Living and working in Windhoek placed him more in his element, and he enjoyed his independence. But his desire for autonomy changed the day he ran into Melanie, whom he had first met in school. She says, "I'd had my eye on him since eleventh grade. I fell in love with him from the first moment I saw him at school, but I didn't have a chance then. Ten years later, though, when I was more confident, my chance came."

Marieta was puzzled by this young woman that Nico brought home, and—as with Jo—she didn't try to hide the controlled chaos that enlivens Harnas. Marieta sensed in Melanie strength and independence, but also eccentricity. Melanie showed up at Harnas for her first visit wearing a pair of pajama pants—not exactly the norm for the van der Merwes. And like Jo's first experience at Harnas, Melanie's first visit was not entirely what she expected.

Nico, Melanie, and her toddler Morgan (from a first marriage) arrived and walked across the bridge onto the lawn. Marieta was outside working and she had a baboon with her—Nick's favorite, Tommy. Tommy immediately jumped on little Morgan and bit her.

Melanie remembers, "All hell broke loose because Nico wanted to protect Morgan and Marieta was trying to get to Tommy to protect him. What an intro! And then, of course, the first meal at the van der Merwe house was an experience. Meals are always an adventure there. Between the baboons trying to steal the food and jumping all over, I felt like I was in the twilight zone. Never a dull moment!"

Although Melanie turned down Nico and Marieta's offer to take her on a lion walk the next day, her love for Nico was more powerful than her terror of baboons and lions, and she eventually learned to tolerate the wild animals on Harnas, hiding her fears and learning to be brave. Nico's complete acceptance and love of Morgan helped win her over. From the beginning, he thought of her as his daughter and she thought of him as her father. Morgan still claims that she inherited Nico's brown eyes, his big feet, and his talent in karate. Marieta, Nick, and the rest of the family also accepted and loved Morgan easily. She fit in immediately and became as much an animal lover as any van der Merwe.

In 2001 the threesome moved to Harnas to manage the restaurant there, staying for a year before moving back to Windhoek to work for a company called Wilderness Safaris, which sets up and guides groups of tourists on adventures in the Namibian bush, a job Nico was well qualified for. Even after Melanie found herself pregnant, she resisted marriage, but finally she gave in when Nico argued that he wanted his new baby to have the van der Merwe name. Five months pregnant with daughter Nica, barefoot on the beach at Swakopmund, the couple was married on November 8, 2003. When Nica was born, Nico was ecstatic: two beautiful girls to love! Morgan had entered school in Windhoek, so she was busy most days, but Nica went everywhere with Nico. She was a beautiful baby, and other than her blue eyes, she was a miniature version of Nico in personality and character—gentle and shy, but with a beautiful smile and a surprising sense of humor—much like her grandfather Nick as well.

Nico took over responsibility for Nica as much as possible, and when she was 18 months old, he started carrying her on his back during safaris into the wilderness. Father and daughter grew very close. She played being his co-driver on a tractor or in an airplane. At night—long after Nica had gone to bed—he sometimes couldn't resist waking her up to watch the National Geographic or Discovery channels with him. In return, Nica became devoted to her father, watching and imitating his every move.

When the family visited Harnas, the two little girls took to the animals naturally. Morgan played with the Bushman children much as Schalk and Marlice had, ending up with all the childhood diseases that come from playing in the dirt and around animals; she even had head lice and ringworm.

Melanie doesn't love animals the way her husband always had, but she, like Jo, brought a new dimension to the family that helps maintain a balance of temperament. The family continued to live in Windhoek and visit Harnas whenever they could. Several times they moved to Harnas and then back again to Windhoek as their careers and hearts led them. No matter where they lived, Harnas was a home they could always return to.

Marlice's future spouse, Rudie van Vuuren, was the only one of the three children-in-law who wasn't intimidated by Harnas, and the only one that Marieta and Nick knew before he came to visit as their daughter's boyfriend. Rudie was one of Schalk's rugby teammates who occasionally visited the farm on weekends. On one of those weekends Marlice was home and she and Rudie met. At the time, though, she was dating someone else.

She was 23 years old, and she had dated a number of young men. All of these relationships had seemed promising, but the test was always bringing a new friend to Harnas and introducing him to the baboons. If the man didn't like animals generally and baboons specifically—or if the baboons didn't like him—Marlice believed something was wrong, and she would end the relationship. One man she dated had real potential, she remembers. "He was handsome, smart, and funny, but when I brought him to Harnas, I was so disappointed! He couldn't drive a tractor, he couldn't ride a horse," and when her beloved baboons jumped on him he said, "'Oh, no. They're getting me dirty!'"

Her boyfriend at the time she met Rudie was not exactly "Harnas ready" either. He worked in the fashion industry and used to take Marlice to fashion shows. Marlice didn't even wear makeup. Her beauty was natural and she was thin but muscular. She sat at these fashion shows that felt so foreign to her and looked at the "anorexic girls" sitting next to her, and wondered "What am I doing here?" The boyfriend was also a drug user, something Schalk knew. In a desperate attempt to save his little sister from what he believed was real trouble, Schalk offered Rudie two steers if he could steal Marlice away from her boyfriend.

It worked, but Marlice says, "Rudie is still waiting for those cattle."

After many boyfriends who struggled to understand this lion-taming girl, Rudie was completely at ease with who Marlice was, and she recognized his ability to blend with her family during their very first visit to Harnas as a couple. "My parents liked him, the baboons loved him, and he could ride a horse!" She didn't have to take care of him every minute on the farm because, she states, "He was confident enough to let me spend time with my family without feeling threatened." Rudie became a doctor, but he loves animals as well, spending more and more of his time recently learning how his physician skills can translate into veterinary care. "Animals don't intimidate him," Marlice says with a sly smile, "even though he doesn't know as much about them as I do."

In December 2000, Marlice and Rudie were married in a unique ceremony at Harnas. The lapa was only partially built— just the thatched roof with grass below. Guests were encouraged to attend barefoot—both the bride and groom did. Marlice wanted animals to play a big part in the celebration, so the always gentle and popular cheetah, Goeters, participated in the ceremony, three baboons attended the reception, and the wedding cakes were created in the shape of large lion paws. The next day Marlice, in her wedding dress, and Rudie had their wedding photos taken in the bush with Kublai, a grown lion. All the wedding guests were familiar with the animals on Harnas, and they all responded positively to the inclusion of animals in the celebratory weekend.

At every wedding, some things are destined to go wrong, so no one expected perfection, especially with animals involved. As planned, Goeters walked down the aisle before the bride, carrying the wedding rings in a box attached to his collar, but in an uncharacteristically moment of shyness, he ran away with the rings. When the minister reached the part of the ceremony for rings to be exchanged, Marlice abruptly stopped him. "Hang on a minute. I'll be right back." Hitching up her long white gown, the bride chased down the cheetah and brought back the box so she could properly exchange vows with her new husband.

She laughs about the reception, which three of her favorite baboons attended, recalling that "those three were the ones who had the best time at the party." Because baboons are so similar to humans, they often enjoy the same things as we do—playing, laughing, socializing, and drinking—with the same result if overdone. Every time anyone put down a partially filled glass of champagne, one of the baboons would steal and drain it. Before long the baboons were the drunkest ones at the party, and no one could figure out how they were managing it until it was too late.

Although the pictures the next day with Kublai turned out well, the wedding night was, as Marlice put it, "a disaster." During the two months before the wedding, Rudie had selected a site in a remote area of Harnas next to a water hole frequented by a variety of animals. There he built a small house-on-stilts, now called the Tree House. The place had no plumbing, though, so the newlyweds had to use the bush as a bathroom. Oversized concrete stairs led up to the door. "Rudie had an upset stomach," remembers Marlice. "He spent the whole night running down and into the bush" to relieve himself. She laughs at her husband's plight and adds, grinning, "I got a good night's sleep!"

In 2005, a son, Zacheo, was born to Marlice and Rudie, the fifth grandchild for Marieta. From his earliest days, Zacheo was introduced to baboons by his mother, who often sat with both boy and baboon on her lap as if it were natural for a mother to nurture two different species at the same time.

So, by the end of 2000, all three van der Merwe children had found partners for their lives, and the family seemed complete. Marlice lived in Windhoek with Rudie, whose medical practice was there, Nico and Melanie were running the restaurant on Harnas, and Schalk and Jo were living and working on the farm. Some tension arose around Harnas with all these new people finding their places in the structure, everyone trying not to do the wrong thing but at the same time attempting to stake out territory. No matter what the people problems, though, everyone kept the same focus: the animals. All the conflicts they worked through were for the animals, and Marieta never let people forget that.

One month after Marlice and Rudie's wedding, though, everything changed. Nick, the patriarch of this family, strong enough to hold everyone together through rough times, developed a disease so rare, the doctors had trouble figuring out what it was. The man who had survived fighting in an antiterrorist squad, crashing in a burning airplane, and working day to day with some of the most dangerous creatures on earth found himself fighting a foe smaller than a pencil eraser. It became his most perilous battle.

Nick and his favorite baboon, Tommy

chapter ten

THE SMALLEST *of* FOES

All sorrows can be borne if you put them into a story
or tell a story about them.

—Isak Dinesen

SATURDAY MORNING, JANUARY 13, 2001, STARTED LIKE every other day on Harnas. Animals were fed, enclosures were cleaned, fences were checked and repaired, and workers and volunteers followed through on their assignments. It was also the day that Nick had promised to do a favor for two local farmers by transporting their cattle to auction in his truck. So many head needed to be transported that every truck in the area was in use. Around 70 percent of Namibians make their living through agriculture, and Nick understood that his overworked neighbors needed his help during auction days.

January in Namibia is hot and dusty. No clouds appeared in the sky to block out the sun from raising the temperature to over 100 degrees. January is part of the rainy season, but the term is a misnomer. When rain does come, it comes in a sudden, heavy downpour and ends quickly. The water soaks so rapidly into the arid soil that an hour later you can't tell it has rained at all. When the breeze blows, it is less a relief than a bother because of the dust it stirs up.

The auction was located in Hereroland, a tribal area northeast of Harnas. The Herero lands cover a large section of northeastern Namibia, south of Bushmanland, jutting up against the Botswana border. Back and forth Nick drove, loading, driving, unloading. Moving cattle is dirty business; by the end of the first run, his face was obscured with dust, and it was hard to tell what lay beneath the layer of dirt. Nick frequently wiped his face with a handkerchief and chugged water all day, but the dirt continued to coat his face, neck, arms, and legs. Each load of cattle had to be lined up, pushed, pulled, and prodded into the truck, while the animals moaned their discontent at being moved from their comfortable grassland. Once at the auction grounds, the process had to be reversed. Then Nick would jump back into the truck, and wipe off what dirt he could out of his eyes and drive back to fetch another load. When he finally got home after the long day, he was exhausted and filthy from the work.

Marieta remembers seeing two ticks on Nick's thigh that night after he bathed, and she watched while he nonchalantly pulled them off. Tick bites are fairly common among people who spend time with animals, especially cattle, sheep, and goats—and most people who work closely with these animals check themselves regularly for ticks. An attached tick is usually not cause for alarm because Namibia is considered a low-risk country for most serious tick-borne diseases like Ebola fever, Marburg disease, and Lassa fever. If someone does develop flulike symptoms that don't improve in a few days, doctors usually suspect and treat—in most cases successfully—malaria or tick bite fever, not something more serious or rare.

So neither Nick nor Marieta was worried about the ticks he had removed. Such bites are a regular by-product of farm living around not just large mammals but also lots of creepy crawly residents as well. Neither of them suspected that at least one of the ticks that had bitten Nick was a rare *Hyalomma* tick. It has distinctive brown and white bands on its legs, if you look closely enough—which most people don't. In southern Africa it's called the bont-legged tick—in Afrikaans, *bontpootbosluise*. An adult bont-legged tick feeds on livestock, but its bite doesn't harm them; the livestock are only carriers for a rare but dangerous virus. Only in people does this virus turn deadly. Called Crimean-Congo Hemorrhagic Fever (CCHF), in Africa it's usually just called Congo Fever.

The next day, Nick began feeling sick—nausea, headache, aching muscles—nothing to be too concerned about, perhaps repercussions of the hard work he had put in the day before. He was relatively young, 52, and in good health. He was more than capable of fending off most diseases. Marieta became more worried when he developed a slight fever, but that could just mean a 24-hour virus, and besides, the fever seemed to come and go without any pattern. Marlice and Rudie were staying at Harnas that weekend, and they insisted on driving Nick to the Gobabis hospital, where doctors did tests for the most likely causes of his illness: tick fever and malaria. Tick fever is caused by small bacterium and is treated with antibiotics. Malaria is rarely fatal if treated early. Both tests were negative, so Nick returned to Harnas, and Marlice and Rudie left Monday morning for Windhoek.

They offered to drive him to the hospital in Windhoek just to be sure, but he refused, claiming he felt a bit better.

But Nick wasn't getting better. He developed red spots on his throat and a rash on his skin. Over the next few days, he also felt agitated and sometimes appeared to be confused about where he was and what was happening to him. With his customary stoicism, he refused to go to the doctor because he had been cleared for what were considered the most common illnesses. He remained in bed, however, because he couldn't summon the energy to get up and work. On Wednesday, an increasingly worried Marieta gave him an ultimatum: show he could get out of bed or he would have to go to the hospital. He tried but couldn't get up by himself, so Marieta decided it was time to get him to Windhoek hospital, where a proper diagnosis could be done. By this time, he was also experiencing nosebleeds and blood was seeping from his gums. The disease was moving rapidly through his system.

This was no flu, Marieta knew, although she had no idea what it was. Only later did the ticks he pulled off seem portentous. On Wednesday morning, Schalk helped his weakened father to the car and drove him to the hospital. Marieta had to stay behind to take care of the animals, and when Nick was driven away, she had no way of knowing it would be the last time she would see her husband alive.

After initial blood tests at the hospital, Schalk and Nick went to Marlice and Rudie's house to wait for the results. In just the few days since she'd seen her father, Marlice noted how much worse he looked: "white, white, white," she remembers. Rudie, who had yet to call his father-in-law "Dad," arrived home with the results of the tests. He walked into the room and his first word was "Dad . . . ."

Nick started at the name. "You almost made me fall off the couch!" he said, laughing.

"Dad," Rudie went on without smiling, "you've got Congo Fever."

"But that's a deadly disease!"

"Yes, it is."

Rudie left the room to regain his composure. He made a vow right then that he would never again treat a family member. He had been married only a month, and he found himself having to tell his new father-in-law that he was most likely dying.

He went back into the room and started to explain the disease.

Nick brushed away his explanation with his hand. "I know what Congo Fever is." He added with his characteristic wit, "Just promise me you won't let the nurses bathe me, like they had to do after the plane crash. I hated having them wash my privates!"

Before they left for the hospital, Schalk helped his father to the bathroom and both of them saw the blood in his urine. They knew it made the diagnosis indisputable.

Rudie, Schalk, and Marlice drove with Nick to the hospital, and then Rudie officially admitted him. Doctors had decided to quarantine Nick, using the only ward they had available for that purpose—a ward formerly used by the army. It had been locked for years, but they opened the doors and Nick became its only tenant. More test results started to come back from the lab, and they were alarming. Nick's white blood cell count was practically nonexistent. When Schalk left his father for the final time, he tried to be positive, telling Nick, "This stay in the hospital will last only for a few days, and then I'll come and drive you home." At heart, however, both men were pessimistic about the outcome.

By this time Nick's abdomen was probably swollen and tender, as the liver is affected quickly with CCHF. The virus was making its way through his system fast. Small blood clots developed in the blood vessels throughout his body, a process called "disseminated intravascular coagulation" or DIC, causing bleeding from his skin, his digestive tract, his respiratory tract, and any open wound—even a small needle mark. Nick's body became bruised because the blood at the injection sites for the giving of fluids and the taking of blood wouldn't coagulate. The blood flowed freely both inside and outside his body from these small needle pricks. Nick's initial confusion and agitation—as is the case with most victims of CCHF—was replaced by sleepiness, depression, and indifference to his situation, changes probably due to lack of proper blood flow.

Inside his body, circulation to his kidneys and other organs was being disrupted, and he developed a jaundiced color around the bruises under his skin and in the whites of his eyes. His body was trying to fight off the disease with its own antibodies, but this was a losing battle because antibodies couldn't be built up quickly enough. The virus was moving too rapidly within his body. Patients who do recover from CCHF begin to feel better in the second week of the virus—usually after day ten. Nick, however, wouldn't make it past day seven of the virus. He would be one of the 30 to 50 percent who do not survive the disease.

No vaccine exists for Congo Fever. Because it is so rare, finding a cure hasn't been a priority, especially in a continent ravaged by poverty and AIDS. The disease is probably much more prevalent than most doctors believe, though. Robert Swanepoel, one of the leading researchers into CCHF, who is working in South Africa, claims that it is "far from rare." He goes on to state:

"I am convinced that the disease is grossly underdiagnosed—it has by far the greatest geographic distribution of any of the so-called viral hemorrhagic fevers—and clinicians simply do not recognize it and so give the patients other labels—somehow it has not caught the imagination in the same way as Ebola and Marburg viruses, which have actually killed far fewer patients to our knowledge."

In the 20 years before Nick's infection, only ten cases had been found and labeled CCHF in Namibia, and seven of those people recovered on their own. But many of the victims of CCHF could have been living in the bush or in remote villages where going to a doctor is not an option. These numbers are, therefore, not included in statistical reports, so no one really knows the extent of CCHF. And while some research on the virus is being done, most of the efforts focus on preventing the infection with the use of insect repellents containing DEET and wearing protective clothing and gloves when working with animals. However, many people living in remote areas lack access even to these protective agents, and the excessive heat in Africa prohibits the wearing of long sleeves and gloves.

As soon the blood test done in Johannesburg confirmed that Nick had indeed contracted Congo Fever, the Ministry of Health took action. They sent a team of health workers to Harnas, to contain the virus by putting the farm under quarantine. Everyone at Harnas—workers, volunteers, guests—had to stay on the farm and the gates were locked. Police officers were even posted to make sure no one got in or out without permission, despite the fact that none of the other family members had shown any signs of the disease. Frantic parents from all over the world called about their children who were volunteering at Harnas. Understandably, no amount of reassurance could calm them since many parents had never even heard of Congo Fever and didn't understand how it was contracted.

The volunteers had had no contact with Nick after his trips to the cattle auction and were not at risk, because direct contact of body fluids is necessary for the virus to be transmitted from human to human. Health workers in protective clothing went into all the buildings at Harnas and, Marieta recalls, "sprayed some kind of powder all over the bloody place to help protect us." Various workers and volunteers on the farm tried to assuage the panic of the volunteers' parents and workers' families by releasing simple statements to the press:

"All of us are very well here."

"Not one of us is sick and we are not even worried."

"We are so calm."

"We did not have contact with our boss, you know."

"You don't go around kissing your boss. You just get instructions to work."

Even though Marlice lived in Windhoek, she wasn't allowed to see her father, but she sent him his favorite ice cream through her husband. Rudie was allowed limited contact to give information to the family, but only because he had been the admitting physician. Otherwise, Nick was alone in his hospital room in the quarantine ward. Only four nurses had access to him.

None of the other family members had shown any signs of the disease. At the hospital, though, Nick's symptoms were rapidly getting worse. He was sliding into complete organ failure. The next predicted steps were inflammation of the lungs, bleeding from the brain, then death from heart failure as the final act of the virus. The doctors became desperate and decided to explore nontraditional cures. One experimental treatment is called intravenous immunoglobulin or IVIG. Because people who survive Congo Fever cannot get the disease again, the theory is that antibodies from the blood of these survivors can help build up the antibodies of those who are currently suffering from the virus. The concentrated antibodies in the gamma globulin could attack the virus, helping the patient to fight it off.

The World Health Organization has made a statement about this treatment, cautiously claiming that "the value of immune plasma from recovered patients for therapeutic purposes has not been demonstrated, although it has been employed on several occasions." Another group working on this therapy is the Virginia Bioinformatics Institute. Their website claims, "There was an early recognition of the possible benefits of treatments using serum prepared from the blood of recovered CCHF patients." The website goes on to say cautiously that "limited studies suggest that CCHF immune serum is beneficial when administered intravenously in [specific doses] over 1 to 2 hours on successive days, and when given early in infection."

At the Windhoek hospital, the medical team decided to try the IVIG process, and a call went out to a known survivor of CCHF. The man lived in a remote region, but he immediately left for the hospital to donate the needed blood.

The family waited. The workers, volunteers, and guests at Harnas waited. All of Namibia waited, actually, since the virus—so deadly and so feared—had made the national newspaper and caused a minor panic. Much of the population misunderstood the way the virus is transmitted, and many feared an epidemic. The Ministry of Health Permanent Secretary tried to calm fears of Namibians, but ignorance of the disease was stronger than the assurance people were getting from the government.

People isolated themselves and stayed inside their houses in heavily populated areas. Fear prevailed.

Nick was suffering greatly, hemorrhaging—blood coming from his gums, his skin, even his eyes. By Friday, January 19, less than one week from the time he had been bitten by the tick, he was slipping in and out of consciousness. Clearly, the gamma globulin was not going to arrive in time to help him. It was Friday, January 19, less than one week from the time he had been bitten by the tick. Nick was most likely unaware of what was happening in his last hours of struggle, and doctors simply made him as comfortable as they could. Rudie remembers Nick hallucinating that Friday morning, talking to his favorite baboon, Tommy, as if he were in the room with him.

It was raining at Harnas when the family received word that Nick had died in the middle of the afternoon.

On Saturday Marieta started to feel ill—sore muscles and a temperature—so they wasted no time in rushing her to the hospital. She was in so much pain by the time she was admitted that no one could touch her. Everyone was filled with dread. The minute she was positively diagnosed—even without the official blood test—the whole family at Harnas was brought to Windhoek and quarantined at a house Rudie rented for them. Food was brought to the quarantined family, left at the door, and the dishes were picked up later. Health workers in special clothing— "like the clothing they use in nuclear places or like a space suit," remembers Melanie—came to the house every day and took blood from the family to keep checking for new cases. No one was willing to gamble on this health scare, which could turn into an epidemic. In Africa, the words "Congo Fever" elicit the same panic and fear as does "E. coli" or "anthrax" in America.

A few days later, Kasoepie developed the same symptoms Marieta was experiencing. She was also admitted and placed in the room next to her longtime friend and boss. A little later, when a nurse went to check on Kasoepie, she found her bed empty. She was in Marieta's room, taking care of her, her own symptoms gone. Were Kasoepie's symptoms feigned so she could slip into the quarantined ward to attend Marieta? Were her symptoms real but psychosomatic, brought on by her closeness to Marieta throughout the years? The answers don't matter. Kasoepie's loyalty to and love for Marieta—despite their sometimes fiery quarrels—guided her actions.

Marieta was given the gamma globulin of the survivor blood and eventually began to get better—whether from the experimental procedure or from her own strong antibodies, and despite the shock that her husband had succumbed to the virus. The quarantine was lifted after about a week, and the family was released, and suddenly life was back to normal—but not normal at all. Marieta was beyond consolation. She describes herself as "completely out of it." She was so afraid of running Harnas by herself that she admits she became "a real bitch." In addition to the emotional shock, for months Marieta continued to suffer pain in one leg as a result of the infection.

Nick had to be cremated by government edict because of the infectious nature of CCHF. His family held a funeral at Harnas where family, neighbors, and many important people of Namibia came to pay their respects. Nico, Schalk, and Marlice each expressed their thoughts. Then Nico and Alwyn Bierman, their pilot friend from the plane crash all three men had survived, climbed into a small plane and flew up into the late January sky, where Nico cast his father's ashes over the land and creatures he loved, worked with, and was taken by.

*Teacher Mara Kuhn leads the Bushmen children in songs
at the Cheeky Cheetah Day Care Centre*

{ chapter eleven }

THE EVOLUTION *of* HARNAS

*If you have men who will exclude any of God's creatures
from the shelter of compassion and pity, you will have men
who will deal likewise with their fellow men.*

—St. Francis of Assisi

IN THE AFTERMATH OF NICK'S DEATH, EVERYONE IN THE
family was scared. They all felt they needed to take control, to
study Nick's books and figure out his systems, and to discern the
future of Harnas. Marieta, in her shock and grief, still recovering
from her own serious bout with Congo Fever, felt that everything—including herself—was falling apart. She felt particular
stress because she realized how little she knew about running the
business end of the farm.

Generally, a lone woman in a patriarchal society like Namibia's is perceived as weak and vulnerable, but most people didn't
know how tough Marieta is. She may have been in shock, she
may have been temporarily confused about her future, and she
may have begun to question her strength, but her upbringing

and her sheer strength and determination saved her. Inside, she was still that young girl who had taken care of her mother in death, stood up to her teachers in school, and boxed with boys much larger than herself.

Looking back to that time, Marieta remembers the vultures descending. "The day of Nick's funeral, a German pilot and his friend flew in and said they had come to buy Harnas because, they said, they knew we were bankrupt. I didn't know if they were telling me the truth. They claimed they wanted to save the animals by buying Harnas. And all these things made me so afraid because I didn't know what was going on in the books of Harnas. I knew nothing. I had never worked with those things; Nick did everything like that. I was so nervous. So I fought and fought and told everyone we must be careful with money and not go into overdraft. Everything must work out and be all right." Eventually her common sense prevailed. She told the German pilot to leave, in no uncertain terms, a command probably peppered with some of her colorful language. He did not return to make the offer again.

Marieta moved out of the house she had shared with Nick. She slept here and there on the farm, trying to avoid painful memories. At last, after two full years, she reopened the house and announced that she was finished mourning. She divided all of the possessions she had shared with Nick into four groups, giving one portion to each of her three children and keeping one for herself. She did some remodeling and created for herself a home where she could be, if not completely happy, at least content. Marieta knew that Nick would have wanted her to continue with her dream, even though she felt less solid, less sure of herself, than she had with Nick as her partner.

As Marieta struggled over learning the books and making sure the animals had what they needed, everyone else watched and tried not to make mistakes, as they themselves grieved for Nick and tried to reimagine their lives without his strong presence. Schalk, Jo, Nico, and Melanie worked together, but they fought together, too. As the months passed the situation became so stressful that Nico and Melanie decided in October that things on Harnas were stable enough and it was time for them to leave. They moved back to Windhoek. On one hand, Marieta hated to see them go because it meant her family would shrink again, but she admits, "I was ready. I knew the books. And we weren't bankrupt. I needed my children after Nick died, and it was lucky they were with me." But as always, Marieta was the real strength, and no one coming or going would change that. She was not only the cement holding Harnas together, she was the heart and soul of the place as well.

The question now was what to do with Harnas. With Schalk and Jo at her side, Marieta needed to decide the farm's future direction. How much of a family operation could it remain? How much of the outside world could she let in and still love each individual animal she rescued and raised? Marieta constantly worries that she is losing touch with the creatures she loves, as volunteers take more responsibility for the animals and more of the daily contact with them. Still, she knows that compromises must be made.

All the while the animals kept arriving, animals that needed help even more than Marieta did. "The good thing is," she states, "as more animals arrived that need healing, the more healed I became. When I felt sad, instead of crying about the loss of Nick, I went out to sit with my baboons or took the lion cubs for a walk.

I still do. I sit with them, talk to them, coo to them—and when I leave them, somehow I feel refreshed, healed, and ready to take on my responsibilities again."

After Nick's death, Marieta began to see an evolution taking place on Harnas, and she began to visualize a new kind of Harnas—where rescuing animals was only part of her mission. People needed to see how saving wildlife benefited them all. Education of tourists and locals alike needed to play a major role in her work. Wherever Marieta went in Namibia she often carried one of her baboon or monkey infants with her, so teaching people came naturally to the situation. Having a baby baboon in a public place draws tourists and natives alike to her, many who want to touch and coo over the sweet tiny almost-human face and hands of the animal. Marieta always uses these encounters to educate people, reminding them that baboons, cute little creatures when young, will grow up to become huge and usually fierce, with powerful canines capable of doing great damage. She also makes it clear that she doesn't *want* to do this—that animals should be free—but because baboons are allowed to be shot as "pest" animals, someone must care for and raise the orphans.

A second way to educate tourists is to bring them to Harnas, where she can teach them while fascinating them with the animals. This meant she needed to have a place that could compete in the growing tourist market in Namibia. The guest area on Harnas needed improvement. Marieta wanted it to be more attractive to people so they would be willing to drive the three hours from Windhoek, although there are game parks much closer to the airport and major cities. The lapa, the lawn, a pool, a volleyball court, outdoor benches and chairs, and of course, guesthouses had to be built or improved.

Eventually, what Marieta created was a beautiful, authentically African setting where people and animals intermingle freely, and guests become part of the Harnas family of creatures and humans. The lapa was built across from the complex that houses the family members. This large indoor/outdoor building with a high thatched roof is furnished with long tables for guest meals or for group meetings. Rich-colored couches strewn with contrasting throws and pillows fill another part of the lapa, and a crescent-shaped bar with seating both inside and out completes the guest area. Outside, next to the bar, wooden tables have been placed under a camelthorn tree with so many spreading branches, you can get dizzy following their paths. And if you do look up, you might very well see three or four of the Harnas cats walking the tightrope of the tree's limbs.

Everything is decorated with African art—much of it locally created by the Herero, Nama, and San people. A large kitchen, hidden behind the back wall of the eating area, allows chefs to create local and regional dishes to serve the guests, and an office is set off next to the kitchen. Behind the lapa is a fire pit with logs set in a circle. Barbecues—known as *braais* in Namibia—are held here on special nights.

Two donated stuffed animals adorn the couch area—neither of which was raised by Marieta. "That would be like stuffing one of my own children to do that!" she exclaims. A wild lion in a position of ferocious attack guards the office, and visitors often enjoy comparing the size of their own hand to the lion's paw, or marveling at the size of the beast's fangs. The other animal is an African wild dog, a creature few people outside of Africa have ever heard of and one that even many Africans have never seen. These two creatures, frozen in time, elicit visitors' constant interest, and children love to pet their heads like big dogs that are finally tame enough to approach. More important, the two are provided for educational reasons—they teach people about the fragile existence of their populations.

Once the lapa was finished, Marieta and her family began to fill in the area between it and the family house with a beautifully manicured lawn watered with gray water to conserve the potable water in this near-desert. The lawn features rock-lined paths, a small pool with two wooden umbrellas and lounge chairs, various African trees and bushes, a small cactus garden, several fish ponds, and benches. What impresses visitors the most about the lawn, however, is the wildlife that abounds. At various times you might see a giraffe, several springboks or blesboks, tortoises, colonies of mongooses, as wells as dogs, cats, warthogs, donkeys, lambs, goats, pigs, and ostriches. Sometimes lion cubs are brought out to play under the thorny trees, and at other times baby baboons swing from branches. To walk out on this lawn is to walk into animal paradise where the creatures engage with each other and rarely seem to understand how amazing it is that they generally all get along. Sometimes small skirmishes between dogs and mongooses break out, but these are brief, territory is reestablished, and peace returns to the lawn once more.

Along one side of the lapa is a long enclosure mostly for adolescent baboons. Their large space is filled with swings and ladders, trees to climb and jump from, pools of water to play in, and a variety of toys to tinker with. These half-grown baboons spend their days and nights unwittingly entertaining onlookers as they chase, scream, climb, and swing. If you approach the fence and sit down, a brave baboon might come and sit across from you. If you move too close, a quick hairy hand might reach through the fence and snatch something of yours that you had no idea would be of interest to a baboon—a shoelace, a Velcro-banded watch, a bracelet, sunglasses.

Signs on the fence remind guests that baboons are fast and curious, but until you see them for yourself, it is hard to imagine just how quick they can be. If you stick your tongue out over and over, you might get a similar response from the baboon. This tongue-thrusting is a sign of friendship, a communication of trust. After this, the baboon might turn around and show you his backside. This is also a sign of friendship, and you should feel honored rather than offended by the sight.

Continuing around the nearly circular lawn, other small enclosures hold younger baboons, meerkats, lion cubs, or any number of animals. On the other side of the family house is a gated enclosure for trucks and a work area, and beyond that stretches a long open area planted with higher, natural grasses. If you don't look closely, you might miss that this is an animal cemetery, with graves marked by rocks and small crosses showing names of animals who once lived at Harnas. Volunteers who have come to the sanctuary more than once often visit the cemetery, looking for the graves of animals that have been lost since their last visit. It is not unusual to find someone crying over the large grave holding the lion Savanna or the small, seemingly insignificant grave containing the body of a beloved meerkat that was killed by a snake. Grief has no size requirement.

Next to the graveyard, and still bordering the lawn, is the crocodile enclosure, a semi-small pond crossed by the bridge that leads to the guesthouses. The crocs, Ina and Claus, spend most of their days here, lounging in the sun, lying in the grass, or cooling down in the water. When guests cross the bridge, many don't even notice the large sunning reptiles, so carefully have they camouflaged themselves. Once in a while, though, they make their presence known. In November 2008, around dawn, an apparent domestic dispute erupted between the croc pair, and Ina decided they needed some time apart. It's likely that this large male did a version of the "high walk" of crocodiles, lifting his body in the middle with his powerful tail, then "crawling" up the fence made of a series of upright cut logs, flopping and seesawing his way to freedom.

It must have been quite a sight in the early morning to see the massive creature escape and begin his journey across the grass toward the guest swimming pool. He might have been eyeing that turquoise oasis for some time, thinking how superior that water looked compared to that in his pond. He short-armed his way across the 50 yards and slipped into the pool, hardly making a splash.

Just after sunrise, Petrus arrived to do his chores. Petrus is San and at night works as the bartender at the lapa, but in the morning he has other jobs, one of which is making sure the pool is sparkling clean. As he approached the pool, he saw what he thought was one of the slatted, wooden lounge chairs floating in the water. When he reached the pool's edge, he leaned over and stretched out his arm to grab the chair and drag it out—when suddenly the chair altered its shape and became a creature his mind struggled to recognize: a crocodile, partly submerged, its eyes and snout floating just above the waterline.

Petrus made a frantic call to the kitchen crew on his radio: "There's a crocodile in the pool!" and heard laughter on the other end. He repeated his message: "I'm not kidding! There's a crocodile in the pool!" The silence at the other end of the radio was followed by the kitchen door banging open. People poured out and ran to the edge of the pool—before backing up a few feet.

Every radio in the place went wild. Most calls went to Frikkie and Schalk—the two men who most resemble crocodile hunters at Harnas. Within minutes everyone gathered at the pool, shaking their heads in amazement at the crocodile that filled the small pool nearly from end to end. Ina's head was at the shallow end by the stairs, which helped keep him from disappearing too deep into the water. He didn't seem bothered by all the attention he had aroused. A debate began about the best way to engineer the rescue and return Ina to his enclosure. Frikkie ordered volunteers to gather nets, sticks, and ropes.

Eventually Frikkie and Schalk created a noose held between two sticks and they carefully dropped it lower and lower until they caught Ina's jaws, holding them closed. Ina came to life, doing the crocodile roll that is a staple of all jungle movies. Frikkie and Schalk held on while Ina wrapped himself tighter and tighter in the noose. Finally, when he slowed, they threw a towel over his eyes to calm him, threaded a net under him, and began to pull him out of the swimming pool and up the stairs. The volunteers jumped into action at Frikkie's command—some of them getting into the pool next to the calmed Ina to hold his legs back against his body. Up he came, passed hand-over-hand from one volunteer to another. The caravan of people carried Ina back to his home, passed him over the fence to other volunteers waiting, and the process of tangling was put in reverse. Finally, Schalk pulled off the towel and jumped over the fence before Ina's thrashing jaws could catch him.

Congratulations were passed all around, everyone agreeing that it had been a once-in-a-lifetime experience.

But they were wrong.

Three days later, a tourist walked over the bridge in the afternoon and was greeted by the sight of a nine-foot crocodile sunning himself on the grass. The uproar started all over again, but this time the rescuers had several disadvantages. First, Ina was not down in the pool, trapped between edges. He could move around freely—and that included his snapping jaws. Second, Ina knew what was coming and was not pleased with the prospect. This time he was much more aggressive and bad-tempered.

The same method was employed, but the noose was not as tight because Ina didn't spin. Eventually Schalk and Frikkie had to jump on the croc's back to hold him in place while volunteers attempted to pick up his thrashing body. Jo stood nearby, watching her husband wrestle the crocodile. It's probably safe to say that most wives in her position would be wincing, praying, and begging their husbands to be careful—or maybe calling on someone else to do the job.

But not Jo, who has become significantly braver and tougher in her years at Harnas. She yelled from the sidelines, Schalk's own personal cheerleader. "Come on, Schalk! You can do it! You've wrestled lions and leopards and played rugby in the World Cup! This is nothing! Get in there and grab that thing!" Schalk—as he is apt to do—smiled, shook his head, and said nothing.

Ina was finally wrangled and returned unwillingly to his enclosure, and Marieta spent the rest of the day redesigning the fence security, adding electricity so the lurching crocodile could never get out again.

Next to the crocs are the first of the outer enclosures that house the older, more self-sufficient animals. The closest one, home to five semi-wild cheetahs, is one of the first enclosures guests see when they enter Harnas and head to the office to check in. Surprise and delight show immediately on all their faces—along with disbelief and maybe fear—as they come face to face, only two feet away, with five sinuous full-grown cheetahs that saunter to the fence to catch a glimpse of the new arrivals.

Across the bridge over the crocodile enclosure, through a sliding grated fence, Marieta placed the guesthouses and camping sites, beyond the safety of the lawn area but still protected from the enclosures for lions, larger baboons, wild dogs, leopards, and cheetahs. The cabins and bungalows are close enough, though, for guests to be treated to the sounds of lions roaring and baboons screaming, especially at night. Guests also have frequent encounters with the less dangerous animals that roam this area freely: ostriches, oryx, warthogs, and small antelope-type creatures like springbok, duikers, and kudu. Visitors may get a peek at the wild giraffes and their offspring or even a zebra. If guests leave the door to their guesthouse open in the afternoon, a colony of mongooses or a mob of meerkats might visit—always a good way to clear the living quarters of pesky bugs, scorpions, or the rare snake, as snakes are well-known enemies of these small but fierce creatures, and scorpions are a favorite food source.

Once Marieta laid out these physical areas, Harnas became much more competitive in the tourist industry, but money is always a concern for a place like Harnas. The constant need for food, volunteer help, and maintenance has kept Marieta focused on the improvement of the sanctuary's financial stability in whatever ways possible. Building on the trust fund created before Nick's death in 1997, in 2000 Harnas was made an official foundation (first in Germany), which made it easier for people to make donations. Recently the foundation has been set up in the United States by a former Harnas volunteer, Catherine Leon, from San Francisco, who wanted to continue helping Harnas long after she left the farm. She has set up the Harnas African Wildlife Rescue in the United States with its own, separate website: harnasusa.org. Although the organization is new, Catherine has already held a fund-raiser in the San Francisco Bay Area and hopes to hold others throughout the U.S. She applied for and received an IRS tax I.D. number so that Americans who want to give money to Harnas can do so as a tax-deductible donation.

In the last decade word about the magic of Harnas has been spreading to people in places near and far away from Namibia. Several South African and German documentaries have been screened, and travel articles have made the rounds in Europe, especially in Germany, but also in the United Kingdom. Most recently, the Animal Planet channel did a series of short films for use between programs, and although they were aired only in the United Kingdom, people in other places can find them on YouTube.com.

Among the many people who have visited Harnas are some who have become special friends and great supporters of Marieta's dream. One of these, Ulla Kaehler, visits Harnas for several months each year, trips she makes possible by living frugally the rest of the year. The two women became good friends because Ulla's love for animals is genuine and fierce, and Marieta noticed that Ulla had a gift for raising baby animals.

Some of Marieta's favorite stories center on Ulla, who started coming to Harnas in 1999. That year, Harnas had adopted an orphaned donkey named Leon, and the two women agreed that

Ulla would become Leon's caretaker. For the rest of Ulla's stay at Harnas, the donkey slept in front of her door at night. Once, during an afternoon nap, Leon even slept with Ulla—on his back with his legs up and resting against the wall. Leon regarded Ulla as his mother and grew up loving her.

When Ulla sat out on the grass, she honked an authentic donkey bray, and Leon came running to her, jumping on her lap and sitting up with his front hoofs hanging down, like a dog begging for food. As Leon grew and grew, he refused to leave the grassy area and move out to the taller grazing grass with the other donkeys. Pretty soon, he became so large that when Ulla called him, and he came running to sit on her lap, all anyone could see was a donkey with his front legs up. No one could even see Ulla underneath. In all, Ulla has raised seven donkeys, her most recent one a shy male she named Kris Kristofferson.

Ulla has helped raise other orphaned animals as well: three cheetahs, a warthog, a caracal—but perhaps her most successful and amazing parenting involved the baby leopard named Missy Jo. She was born from wild leopards that were caught in a box trap by a local farmer. Harnas took them, of course, but didn't know that the female was pregnant. Some time later, workers found two cubs in the enclosure. The male cub died, and Missy Jo was injured, but she recovered with careful nursing.

Leopards are notoriously difficult to hand-raise because they have an independent streak right from the start that lion and cheetah cubs lack. Also, because large cats often pee indiscriminately, a caretaker cannot bring them into her bed because, more than likely, the caretaker will wake up in a pool of urine, soaked through to the mattress. Ulla attempted a few times to use diapers on Missy Jo, since they'd been used so successfully on baby baboons, but she soon realized that diapers just aren't made for cats, which can claw them off in seconds. Surprisingly, though, Missy Jo showed great restraint while sleeping with her new "mother." Ulla claims Missy Jo never once urinated in the bed, and instead would get up, go into Ulla's shower, and pee there—returning to bed when she was finished!

To this day, Missy Jo remains the tamest of all the leopards on Harnas. Most leopards become aggressive as they age and are untrustworthy in virtually all situations. Missy Jo has been the exception. In her large enclosure next to other, wilder leopards, she waits each morning for her food, but mostly she waits for the people who bring it. She makes a wild sort of meow—almost a "yowl"—common to leopards, and rubs herself on the fence. Brave volunteers reach through and cautiously stroke her thick, velvety pelt, and Missy Jo ignores the food if a human hand is petting her. When Ulla is in town, she can still enter the enclosure and give this most unusual leopard a good scratching.

Another story characterizes this German woman's determination and bravery. Hard as it is to believe, at one time Marieta had 13 baby baboons in her house. At that point she was unwilling to let volunteers nurture them at all, so all 13 slept with her. Marieta's ordinary bedroom would never take the chaos of these sometimes enthusiastic youngsters, so she had a special room made up with two mattresses on the floor and nothing else in the room that baby baboons might think worthy to use as toys in the middle of the night. After feeding her 13 charges, Marieta would retire to her baboon nursery and sleep the night away, covered in furry friends. Ulla, always willing to try something new and adventurous, asked Marieta one night if she could sleep in the baboon nursery. Like a lot of people, Ulla was a little intimidated by baboons, which can be surprisingly strong, curious, and mischievous, but she figured this was the time to get over her fear and have a new adventure.

At bedtime, Marieta gathered up her 13 babies. Some of them she carried in her arms, some sat on her shoulders or head, and others wrapped themselves around various parts of Marieta's legs, and together with Ulla, she walked to the special nursery, lay down, and went to sleep. Marieta woke up in the middle of the night and decided to check on Ulla who, after all, was sleeping with baboons for the first time.

Marieta turned on her flashlight and looked at the other mattress. "There was no Ulla there. Baboons had completely covered her, including her face and head. Ulla was so afraid to move, afraid that a baboon would somehow be hurt or hurt her, that she remained motionless—awake and trying not to be terrified that baboons were sleeping on her face." The night cured her of her fear, but Ulla had had enough. "One night with baboons was all I needed," she says with a grimace.

Such good friends and supporters can never replace the huge hole Nick van der Merwe's death left, yet Marieta continues to move forward and expand the dream. She was determined to make the Harnas mission a success. Nick's legacy is evident all around Harnas, in the home he built, the children he fathered, and the vision he shared with Marieta about saving the wildlife of Namibia. Because of Marieta's belief that life's events have purpose and meaning, she can speak of her husband's death philosophically. "Nick crashed in the plane and survived, but this rare thing—this small tick—came and took Nick. It seems Nick had to die from something rare—not a normal illness or a plane crash. He was special, and his death had to be unusual enough to fit him. Even the newspapers talked about this. I think it's something that had to happen this way."

The author with 3-month-old lion cub Brad and lab Brolloks

chapter twelve

VOLUNTEERS

*I have always held firmly to the thought that each one of us
can do a little to bring some portion of misery to an end.*

—Albert Schweitzer

ONE OF THE MOST IMPORTANT YEARS IN THE EVOLU-
tion of Harnas was 2003, because that was the year that Frikkie
von Solms, Marieta's cousin and friend from childhood, came to
Harnas to retool the volunteer program.

Before that time, the use of volunteers' services had been
inconsistent and disorganized. Enough miscues occurred to convince Marieta and her family that they needed more control over
the young people coming to help. Harnas simply could not add
newcomers ad hoc to the farm's existing routines. For instance,
when Marieta first started accepting volunteers, Jo would usually
drive to Windhoek to pick up the new recruits—even though she
was still a city girl, uncomfortable driving the three hours each
way, mostly through bushland. The third time she made this trip,

disaster struck. Jo picked up four Swedish girls from the airport and began the drive home with them, her infant son, Samar, and Samar's nanny. About a mile from the Harnas gate, the car quit. Sunset was coming on fast, and Jo had no radio to call for help.

She knew she had to portray courage and control for her passengers.

"Girls," Jo announced, "leave your bags in the car. We'll have to walk to Harnas."

What she didn't tell them was that the house was almost seven miles inside the main gate. Samar was hungry and crying, but the six women and one baby began the journey. Jo put on a mask of authority for the volunteers, answering their fears with reassurance.

"No! This is perfectly safe!" she lied. "It's not dangerous! We'll be fine!"

They walked and walked. Samar finally fell asleep as they plodded on. About a half mile from the main house—now shrouded in total darkness—they heard cracking through the trees.

"Stay calm!" Jo ordered. "But get yourself a tree and climb it!"

The girls and the nanny stumbled and struggled but did what Jo told them. She handed Samar up to the nanny and continued with her facade of bravery. She grabbed the biggest fallen tree limb she could find and lifted it in front of her, straining to balance this branch that was practically as big as she was.

"I'll stay down here and guard you," she told the wide-eyed, speechless girls, who were getting their money's worth of African adventure on their first night.

Unbeknownst to Jo, earlier in the day Schalk had released two eland bulls. Elands are the largest of the African antelopes, the males weighing up to a ton and a half and standing up to six feet tall. The adults are blue-gray with vertical white stripes on their sides. On their heads they sport long, spiral horns. They are active at night when they feed—which is why Schalk had released them, not for one minute believing that the women would be walking home.

When the bulls arrived where Jo stood, the only obstacles between them and Jo were a small fence and the tree branch she was clutching. The bulls charged the fence over and over, for more than an hour, and Jo continued to protect the girls, her son, and his nanny in the trees by jabbing the branch and yelling at the huge animals. Normally Jo is terrified of many animals on Harnas, and she admits she could happily spend her life on the grassy area between her house and the lapa. That night, however, she was fierce because she had to be.

Marieta realized that Jo and the new volunteers were extremely late, so she knew something was wrong. She sent Schalk in a truck to search for them, and he found their tracks, leading him to the trees where they were under siege. His exhausted wife saw him and screamed, "Save us! Save us!" As the bulls continued to charge the fence, Schalk had to pull the truck right alongside the other side so the women could climb in.

When enough risky situations like this had occurred, Marieta and the family knew decisively that Harnas needed help with its volunteer program. Frikkie seemed the perfect answer to the problem. Not only did his background make him a perfect candidate for the position, but his love for animals set him apart from others who could have created a successful volunteer program.

"You know," Frikkie remarks, "people are a great deal like animals in their behavior. For example, wild dogs are pack animals, and so, generally, are people. Watching how the dogs act and react in their natural setting helps me understand how volunteers work in a group." He takes a moment to consider this comparison as he blows smoke toward the azure sky. "It's an easy transition from animal to human. Animals, like people, have distinct personalities and moods. If more people realized this, I think we'd all get along better."

On the outside, Frikkie is one of those men whom people describe with clichés: tough as nails, . . . as leather, . . . as steel. He is nobody you'd want to tangle with, no one you'd want to cross, no one whose hit list you'd want to top. But Frikkie's personality is much more complex than that. He would give his all to save the life of even one animal.

One afternoon Melanie walked from the family houses to the lapa, carrying her new terrier—a puppy that weighed a mere four pounds but had energy enough for a pack. When Melanie reached the lapa, the tiny dog jumped abruptly and inexplicably from her grasp and landed with an awkward and horrifying thud on the cement. A five-foot fall for a four-pound dog is a bad one, and the dog immediately began seizing. Melanie screamed. People came running when they heard Melanie's cry, and one of the first to arrive was Frikkie. Without hesitation, he bent down to the toy dog and began to give artificial respiration. He continued while the others looked on, helpless, until it became obvious that nothing could save the tiny creature. This wasn't the first time Frikkie had performed artificial respiration on an animal.

He watches the animals carefully as volunteers complete their duties. He expects to be informed of any change in behavior or possibility of injury or illness in an animal. Every life is important to him, and every death is a cause for deep grief, whether from illness, injury, or old age.

As a young man Frikkie had not aspired to caring for Namibia's dwindling wildlife, although he had grown up on a farm and had a thorough knowledge of animals. His parents lived just seven miles from Marieta's, and it was to their home that Marieta had gone after her mother died. Frikkie is more of a brother to her than a cousin, and he has always remained close to Marieta. His background as a teacher and administrator made him a perfect volunteer coordinator because he understands young people and loves to work with them in the outdoors, honing both their personal and survival skills. Early in his career, he had initiated an "Outdoor School" for the government, similar in nature to the Outward Bound program in the United States. He would depart each Friday afternoon with 30 boys or girls to venture into the desert where, in his words, he would "break them down and then build them up."

After many years of service to the education of Namibia's youth, Frikkie had retired, but then the call came from Marieta saying she needed help with the volunteer program. With the number of animals continually rising, the workload was becoming too burdensome for the family and the few regular volunteers, most of whom had learned about the program by word of mouth. Someone had to organize, expand, advertise, and professionalize the volunteer program.

Before Frikkie took the reins, volunteers would show up in Marieta's kitchen at any time during the day and ask Marieta what they should do. Although she always needed help, it was difficult for her to delegate responsibility, because she wasn't able to follow through. So she never knew if the work was actually being done—and done the right way. Before long, she would see the volunteers again, and their questions and problems would come in rapid fire:

"There's no chicken for the cheetahs!"

"How many baby foxes are there? We can only find six."

"Where are the keys to the room with the blankets?"

"I can't find the bowl for Goeters."

"The meat is frozen again and it's impossible to cut!"

"Where's Etosha?"

Marieta usually answered these queries abruptly. She has little patience for people without initiative—who ask questions instead of searching for the answer themselves—or for people who don't take responsibility and make things happen without help. She would wave her hands, pushing the questions and the volunteers away as she returned to her own considerations, saying she would take care of the problem herself. The volunteers would retreat, left uncertain by her rebuff. Despite her unflagging devotion to saving animals, she is not by nature a manager of people, and although the volunteers wanted to help, she knew they lacked the skills that she and her family members had.

In Frikkie she found an ideal director of people. That division of labor has freed her to operate on her own, as she always prefers. For most volunteers, Marieta is a mysterious, awe-inspiring figure, a woman who moves through Harnas fixing, yelling, cussing, often in her own world, sometimes not hearing people when they petition her. Marieta's brusque ways and fleeting encounters with the volunteers only add to the mystique surrounding her inspiring but puzzling persona. Occasionally they'll find her playing with baby baboons on the lawn, and they'll creep up and engage her in conversation carefully, almost reverently.

She is not a maternal figure. Her own upbringing did not lead her to coddle children. She believes people should fight through their difficulties on their own and move forward—and this includes herself, her family, her workers, and the often sheltered and privileged volunteers who come to Harnas from well-to-do families in Europe and North America. They may never have experienced the kind of responsibility and workload facing them at Harnas, but they're expected to adapt quickly. When animals are sick, hurt, or in need, Marieta is the first to help, but people, she says, can "bloody well manage."

Still, she knows that opening up Harnas to others was the right move. Giving up a good chunk of the individual nurturing of animals was difficult for Marieta, but looking back, she sees that having many nurturers for the animals has turned out to be a good thing for them. "My animals are better than before, when it was only the family working with them. Now it's Mikkel and Amy and Ashley and Matt and so many different people. The animals aren't afraid of new people. They are tamer."

Frikkie did not immediately decide to join the van der Merwes. Taking the job would mean living apart from his wife a good portion of the time because she was a teacher in Windhoek. He would be joining a family business, and with that, he knew, would come problems. He thought about his options, the challenges, and especially the animals, and came to the decision to join Harnas, but on a consulting basis only—no salary, just a daily fee.

"I was in a permanent job for so long," he says, "and I wanted a bit of movement. I find if you become one of the pack, you are not observant anymore and so lose perspective. I can be more objective this way—even though I'm actually family. You can't be part of your team. You have to be in front of it. People in the family can influence you and then you become emotional, and that's when you make mistakes. You need to see reality—the facts.

You cannot disagree with the family without feeling so, well, aaaagggggghhh!" He looks away and shakes his head. "As a consultant, I have more freedom and independence."

Frikkie schedules his work hours to correspond to the times when the van der Merwe family needs him the most. Schalk has the most responsibility these days, Frikkie observes, so "I try to be there for him when he needs to talk." Marieta can also feel free to talk to Frikkie when she doesn't want to talk to the rest of the family. He becomes a sounding board for everyone, and because he knows so much about so many things—"jack of all trades, master of none," he claims—he can give excellent advice on a variety of subjects from tourism, finances, volunteers with emotional problems, family disputes—whatever is needed.

The Harnas volunteer program, also called the Working Guest Program, has evolved into a well-organized machine under Frikkie's direction. Instead of trickling in whenever they please, incoming groups of volunteers arrive on Friday of each week and departing groups leave on Thursdays. They are driven by a professional transport company, to avoid situations like the one that found Jo wielding a tree branch to protect volunteers from eland bulls. This organization gives Frikkie the opportunity to give a good orientation to each group arriving, and also to talk to departing volunteers who will be resurfacing in the real world again, also as a group.

Some volunteers stay for two weeks, and some for as long as three months, but no matter the time frame, the first week is the most important, he explains: "They come here with certain perceptions that may or may not be true. They need to have a good understanding of Harnas, the animals, and the organization—and that means discipline, something many of these young people don't have."

Most volunteers are between the ages of 18 and 25, although more and more volunteers in their late 20s and 30s have been drawn to Harnas—and even a few in their 40s. The largest number comes from the United Kingdom and other parts of Europe. Young people in Europe often take a "Gap Year," during which they travel, volunteer, and get a better global perspective, and many volunteers come to Harnas as part of that experience. Organizations are set up to help these seekers find the adventures they want, and the ones who come to Harnas come generally because they love animals—although they're never really prepared for the kind of work they'll have to do. Among the volunteers, the ratio of females to males is around eight to one.

They live in cabins in groups of four, and they sometimes have only cold showers, depending on the season and whether their energy level is high enough after a long day of chores to build a fire under the water heater. They experience occasional losses of electricity and do dirty work from sun-up to as long as they can stay awake. They often find themselves covered in animal urine and feces, dirt, and blood. Their arms, legs, and sometimes faces are covered in scrapes, scratches, and bruises. Some of these come from walking through the bush; some from playful attacks by leopards, lions, and cheetahs; and some from opening and closing the large number of gates that separate the animals. (Volunteers often call those gates, with wires sticking out at odd angles, the "most vicious animal on Harnas.") Yet so many volunteers extend their time, realizing that they aren't ready to go back to civilization, aren't ready to leave their animals. They rearrange airline reservations, work schedules, and semester classes just so they can stay and work longer under these unsparing conditions.

Originally, volunteers stayed in the "volunteer house," connected to the family houses, but Frikkie knew this setup had to be changed. It meant no privacy for the volunteers and no privacy for the family. After several years of this too-close intimacy, the Volunteer Village, a ten-minute walk from the family home and the lapa area, was finished in 2007. The cabins and toilets were built around a group area used for meetings, meals, and fun. A small pool—dug by volunteers in 2008—offers relief from the pitiless afternoon sun, and a huge fire pit gives comfort at night. The volunteers are able to bond without worrying if they're keeping the family from sleeping. The van der Merwes sighed with relief when the village opened. Finally, some peace and quiet. The old volunteer house was remodeled as a permanent home for Schalk, Jo, and their children. Everyone breathed more easily with more space.

Every day begins the same way, but then anything can happen. Jo told Animal Planet, "You get up in the morning and you say 'Today we're going to clean all the cribs and put up new enclosures.' Then, half an hour later, the volunteers say, 'This has happened and that has happened,' and we say, 'All righty, then, we'll fix that and do the cribs and enclosures tomorrow.'" Most volunteers get up around 7 a.m. and fix their own breakfast: cereal, toast, coffee, yogurt. Frikkie joins them at 8 a.m. and outlines the day's proposed activities—who will do what extra chores, who will go on the guest tour to feed the animals in the outer enclosures, what larger jobs will need to be done. He then provides various pieces of information about the animals—such as sickness, deaths, or births.

Frikkie has organized the volunteers into four groups, each with its own responsibilities for specific animals. They have group leaders who have been at Harnas the longest, and they take care of small problems within the group, only bringing larger issues to Frikkie's attention.

"It took me about six months to get the program where I wanted it," he observes. "And now I can work on honing it to new levels of efficiency and think about expanding it in new directions."

Each of the groups has a list of animals they must feed and enclosures they must clean, so immediately after the morning meeting most of the volunteers go off to the food-preparation area. Because so many of the animals at Harnas are carnivores, volunteers must get used to handling raw meat. For some this is a shock and they take some time to adjust to the blood and gristle, bones and hide.

Sally, from Birmingham, England, laughs when she describes her "before" and "after" selves. "Before I left home, I was a prissy princess girl, but after my first week, I stopped showering so much and was covered in blood during food prep with my hand up carcasses, pulling out intestines. If someone had told me I'd be doing that before I left home, I wouldn't have believed it. But you adjust so fast." She also believes that she needed to come— and in fact was led to come. "When I got here, it felt like part of me was missing, and I found it here. I think when I leave it will be missing again because it's part of me that belongs here." Another volunteer describes seeing food prepared for carnivores for the first time and smiles, "The first day I walked into food prep and saw a horse's leg, and I thought, 'This is going to be foul,' but actually it was fine, especially when you see the animals eating and you appreciate more and more about survival."

Each animal has its own food bowl with its name or species on it, color-coordinated to the volunteer group responsible for that animal. While the food-prep area might look chaotic with everyone doing something different, knives flashing all around, and foods of various kinds being divided up, the chaos is controlled because everyone knows exactly what he or she must do. New volunteers, after a day of orientation with Frikkie, are led to the food-prep area and told, "Pay attention because tomorrow you'll have to do it yourself." Sure enough, the next day, the new volunteers prepare the food bowls for their animals, but under the careful observation of experienced volunteers. By the next day they are experts, and when new recruits show up the following week, the Rookies become the Veterans, and the whole process starts over. Volunteers—innocent and ignorant when they arrive—become mentors in short order.

The food-prep area clears out as individuals carry bowls of food to every kind of animal imaginable, from the biggest lion to the smallest fish in the pond. No one misses a meal at Harnas. A different group is assigned each day to clean the food-prep area, and by 11 o'clock the tables, bowls, knives, and floors are spotless—not just because it's the right thing to do, but because it keeps down bacteria and disease. Cleanliness is especially necessary when working with raw meat. The volunteers then spread out to clean the enclosures of their animals, raking up excrement, leaves, and fallen food. This also keeps bacteria from forming and certainly makes each enclosure more enjoyable to visit.

A surreal chorus of "Brrrrrr! Brrrrrr! Brrrrrr!" erupts from a group of volunteers as they enter the grass area. This continues for several minutes until they get the response they are waiting for: a colony of mongooses appears from around the corner of the family house and another group from behind the lapa. They swarm, looking like small waves of brownish-gray fur surging across the grass. The volunteers begin throwing out small cubes of meat and the mongooses compete for each chunk, while making their own chirping noises in response to the "Brrrrr! Brrrrr!" sounds the volunteers continue to make. Other volunteers cross the grass with bowls of food for cheetahs and baboons in enclosures behind the lapa, and they laugh at the comic antics of the small creatures. Only one of this colony of mongooses is tame, Collin, and the only way he can be distinguished from his friends is that when a volunteer sits on the grass, Collin will climb up on her lap and start burrowing in.

With startling experiences like having a mongoose make himself comfortable in your lap, feeding the animals is never all work, and various volunteers erupt in laughter as they observe the antics of the animals. You cannot rake a vervet monkey's enclosure without getting jumped on a half dozen times, and the volunteer will gladly put down the rake and spend some time playing with the exuberant monkey. Or in the middle of cleaning out a water hole in a cheetah enclosure, it's hard to resist taking some time to lie down, scratch behind a cheetah's ears, and feel the rumble of the purr in response. Feeding baby baboons practically requires some playtime, throwing them up in the air and catching them while they cackle, swinging them around in circles, and cuddling while baboon-kissing.

One of my favorite tasks was feeding Klippie, the six-month-old baby giraffe. I prepared her three bottles—two-liter Coke bottles, each with a nipple the size of a popsicle—filled with cow's milk still warm from the udder. Feeding a giraffe isn't a skill that school prepares you for, so much of the art evolves from trial, error, and trial again.

The first time I fed her, I was alone, filling in for a more experienced volunteer who was on a bushwalk. I carried the three unwieldy bottles of milk from the kitchen out to the lawn area and called out Klippie's name. She came galloping over when she saw the bottles, a startling sight, since at 5'7", I only came to the top of her legs. She could run right over me without even realizing she'd left my crumpled body in her wake. Klippie was so tall that to feed this baby, I had to stand on a bench and raise my arms, holding one of the bottles as high as possible. Even so, she had to bend a little to get the nipple in her mouth.

Klippie started sucking the milk out with such power that she actually sucked the nipple right off the bottle. I panicked and was afraid she'd choke on it. Could I do the Heimlich maneuver on a giraffe? Yet she worked it around in her mouth like a big stick of gum and then spat it out—about ten feet, I'd say. Quite a spitting queen, that giraffe. I retrieved it and reattached it while she pushed me around impatiently with her nose, almost knocking me over. I climbed back on the bench and held on tight—I wasn't going to let her take that nipple again.

She sucked so hard she took my hand halfway into her mouth, but I held on. She finished one bottle, I threw it down, and we started the second—and then the third. The milk was dribbling out around the nipple and her mouth, running down my arm, into my armpit, and down the inside of my shirt to my waist and toward my legs. When she finished the third bottle, I pulled my hand out of her mouth, and it was dripping with ropes of giraffe drool. I couldn't help grinning at our new bond. Klippie and I nuzzled for a while and I scratched her long neck. Then I headed for the showers.

Another night I was in the bathroom brushing my teeth, and I heard "knock, knock, knock." I thought, "Hmmm. Maybe another volunteer is making a late-night visit?" I went to the outside door and opened it. No one. Thinking that was weird, I went back to the bathroom. "Knock, knock, knock!" I went back to the door, opened it, and again no one was there. Then while I was looking out—wondering whether other volunteers were playing a prank—I heard the knocking again and realized that it was on the window. I went over to the window and opened it. One of the two ostriches—whom I had nicknamed Lucy and Ethyl—was standing there looking at me with her bulbous eyes rimmed with curly lashes. She stuck her head and long, hairy, velvety neck right into my room, backing me up a few paces.

We chatted for a few minutes while I tried to figure out what she wanted. Finally I decided she might be hungry, so I got some of my crackers, opened the door, and went to an outside table and crumbled them up. While I was doing this, the other ostrich joined us, and with one on each side of me, watching me carefully, I sprinkled the crackers and backed away, carefully avoiding their powerful legs, which they use to kick their enemies. They nibbled—and seemed pleased with my gift. They didn't call on me again that night.

Lucy and Ethyl aren't the only visitors volunteers get. If you have food in your room, you have to expect guests, and I was lucky enough to be visited at various times by a donkey named Jasper, Klippie, cats, dogs, a colony of mongooses, a duiker, three oryx, a small herd of springbok, and Murray the warthog.

As surreal as these moments may seem, the encounters do more than amaze and amuse. These individual connections with the animals teach the working guests as much about themselves as they learn about the animals. Jullian, a volunteer in November of 2007, wrote this passage in the online guestbook:

My mother once said to me that life is like a colouring book with a new page to colour in every new day. Here at Harnas, I feel like I've been given a magical colouring book with infinite pages and the most vibrant spectrum of paints to colour every awe-inspiring day. My soul has been laid bare. The routine and materialism that control my daily life back home feel like chains hanging on my heart. Harnas has broken down these emotional barriers and I find myself falling in love with this compelling place. Being here in these magnificent surroundings, I am re-discovering who I am and what is really important to me. Essentially, what makes me truly happy. Through spending precious moments with the animals, I am learning the art of silent communication and embracing the power of mutual trust and respect.

Such wise statements from young people who have spent more than a couple of weeks at Harnas are the rule rather than the exception. The transformations come gradually, but when volunteers look back at the people they were when they arrived, they usually realize how far they've come. One of the experiences all volunteers share is hard physical work—a virtue many modern middle- and upper-class people eschew as beneath their status. But at Harnas, everyone works. After cleaning enclosures and eating lunch, the volunteers find that many afternoons are filled with labor most have never encountered. Sometimes new fences have to be built or old ones repaired, sometimes weeds need to be pulled along electric fences to keep them from shorting out, sometimes an enclosure must be cleared of old bones, and sometimes water holes must be repaired or built. By nightfall, workers are often sunburned and blistered, their muscles sore, but they realize they have accomplished something real and lasting, and that is enormously satisfying.

In 2006 a group of volunteers set out to make a water hole, a task not one of them had ever imagined doing before becoming a volunteer. With Frikkie directing them, they began, and one volunteer recorded her observations:

One day this is going to be one of grandma's old stories around the campfire. . . . We started digging on a hot day around lunch time. Fifteen spades and shovels went with us. We were digging constantly for two hours and we (the 24 of us) took turns in using them. . . . While some people of our group were digging, the rest went to collect stones. They were loaded onto the truck and driven back to the project site. . . . [The truck was] fully loaded with stones and very heavy at the back. It was hard work but we all made it back in time for our late lunch. A couple of days later we went back to the project site. On the way we collected a "bakkie" [truckload] of stones in all shapes and sizes. They were to be used around the bottom of the water hole. We also had to dig out stones from the old water hole at the site. Four people in muddy water up to their knees with loads of bullfrogs jumping all around them did the job nobody else wanted to do. It took most of the day to lay all the stones in the hole, for it to be ready for the cementing. Finally we reached the end! A beautiful name sign was made and there stood our water hole!

The pride of this young woman resounds in the final two sentences. Such physical achievement is rarely possible in the ordinary lives of modern people, and volunteers learn more from these experiences than how to create a good water hole for wild animals. They learn discipline, commitment, and pride in doing a job the right way, without shortcuts. And they bond with other people from all over the world in their quest to do something good for the wildlife of Africa—the real reason most of them have come to Harnas.

Hunter, a young woman from California, knew she wanted to work with animals, and she came to Harnas all by herself—while many of her friends told she was crazy. (Many at Harnas believe that American young adults travel only in packs.) Hunter's transformation occurred gradually over the four weeks she spent at Harnas, and she had several experiences that she believes have changed her forever. One of these key experiences was her work with Boerjtie, the epileptic/Down syndrome baboon. Her work with him, she believes, will directly influence what she planned to do when she got home. She says, "I worked with Boerjtie here, and it taught me compassion and patience. I think I've learned I'm good working with handicapped people, and now I see that much more clearly." Handicapped children, she believes, will be her life's work, thanks to Boerjtie.

Hunter also visited a local school, which helped influence her decision to work with children who are challenged in some way. "I went to the school the other day," she explains, "and I saw children with swollen bellies. It made me think of a few of the volunteers who have been complaining that we aren't getting enough food here at Harnas. When I got back and heard them start up again, I blew up and made a big speech. First of all, we're in Africa where people don't have much, and these four women were complaining! We're getting plenty of food, but we're so used to getting more than we need—too much. I told them to get used to it and stop complaining!"

Hunter went on to describe the effect this continent has had on her personally. "Africa is so raw. It rips you open and you can't get away from yourself. You can't put up a false front. We end up stripped and showing what we're really made of. If you're angry, you show it. If you're sad, you show it. Everything comes out, whether you want it to or not. I feel like I'm naked—but everyone is naked. It's too bad that when I go home, I'll have to put up a wall again in order to survive in that world."

Frikkie tries to help prepare volunteers for returning to the world beyond Harnas. When a group is getting ready to leave, he spends more time with them in their final few days, preparing them for the culture shock they will encounter. The time they have spent on Harnas, though, can be part of what helps them adjust, he explains: "I tell them to take what they have learned about survival and use it to survive in London or wherever they go. The bonding they have created with other volunteers is good, but it's not what lasts. It's the bonding with a specific animal that will endure. It's the animal—and the memory of that relationship—that will help a volunteer survive and move back easily into their reality."

Beyond the shock of going back home, he knows that their families and friends will be shocked, too—because they will no longer really know this changed person who has returned. Everyone will have to make an effort to get to know and respect the new person coming home, and many will resist that because they won't understand how this change came about.

Mikkel from Denmark was a volunteer who kept extending his time at Harnas. He originally signed up for three weeks, but he kept adding weeks until he stayed for eight. At the age of 24, he had finished with his university training and started a new job, but he decided to take a leave without pay to work at Harnas. He had to keep calling his company and begging for more time off, because he felt he simply couldn't leave the animals yet. Three in particular became significant to him: two cheetahs, Pride and Cleo, and the leopard cub named Lost. He slept outside nearly every night in the Namibian winter, when temperatures can drop to freezing. He put down a mattress and a thin sleeping bag with whatever blankets he could find. He didn't have to—he wanted to. He recognized that he had been offered an experience unlike anything he'd ever have again.

The night isn't peaceful when it's spent with Lost, but it is memorable, as Mikkel describes it. "Lost is nocturnal, so she doesn't really sleep at night. She's old enough that she doesn't need me to nurture her at night, but she likes it when I'm there. I've definitely bonded with her. I put down a mattress and a sleeping bag—quickly—because she'll constantly be trying to lick me and jump on my back. She walks around all night long and plays, pouncing on me in the darkness when I least expect it, but I love it. I can sleep when I go home. A night with the cheetahs is different because they'll sleep—really sleep—right on top of me. One of the moments I'll take with me forever is waking up with two cheetahs on top of me. They wake up, stretch, and start to play. That's the image I'll come back to in my mind when I'm back in Denmark and need some peace."

Sally bonded with a different animal: Klippie, that baby giraffe. Sally had been feeding Klippie for about two weeks when she had a seminal experience. "I was sitting on the lawn with Klippie, and she put her head down and started sniffing me. I stroked her under her chin, and suddenly she closed her eyes and sighed this huge, loud satisfied sigh. I had been worrying that I didn't have a connection with her, but that moment changed things completely. I had tears in my eyes. It will stay with me forever."

Eve, also from the UK, became connected to the lion named Hemingway and his mate Simba, but her experience taught her painful lessons about both life and death at Harnas. Eve felt privileged to be selected to feed the lions, and she recognized in Hemingway a lion of great magnificence. He "was the most beautiful and powerful lion you could ever imagine, a beautiful full mane, eyes that lit up like fire, and a growl that would make the ground shudder." Unfortunately, after only a few days, Eve witnessed Hemingway's sudden and sad decline as his kidneys began to fail. She participated in his care and watched while the vet tried to save him. When he died, she and several others felt "totally devastated and inconsolable," sitting by his "cold stiff body, stroking his mane and trying to come to terms with what had happened" for several hours during that cold night.

Although the experience was extremely difficult, she says, "that night gave me some closure on the situation and made me realize just how precious these animals are and how much they mean to Harnas." Eve helped dig his grave, and she spent as much time as she could with Hemingway's mate Simba, grieving together. Eve returned to Harnas for a second time, and as soon as she arrived she "immediately went over to Hemingway's grave . . . remembering the special moments of last year. Seeing Simba again was very special and hearing her roar again was awesome." Eve was also glad to see that Simba had a new mate, Macho, and seemed happy and content again.

Each volunteer's experience is different. Learning about death and participating in the grieving process was important to Eve's personal growth. She says: "Everyone has special moments at

Harnas that will stay with them forever, touching moments that cannot be replaced by anything else in the world. This was my outstanding moment while also the saddest moment I had experienced in a long time."

Most volunteers come to Harnas without a clear understanding of themselves or what they will be doing or learning, and some volunteers are more difficult than others. Frikkie has had to work with volunteers on drugs, delinquents, spoiled and lazy rich kids, and even disturbed youths with suicidal tendencies, habits of cutting themselves or starving themselves. His work with the Outdoor School, though, prepared him for many of these problem young adults. He knows he has to tackle problems immediately rather than wait for them to get worse.

Like the animals that come to Harnas to be healed, people, too, can be healed through their contact with these animals. Frikkie uses his own contact with the animals to survive his difficult job. "When a group isn't working out well and I'm down and I'm frustrated, I'll just go to the lions—play with them for an hour. I love it, and I come back ready to do the job."

Every week or so Frikkie takes the group of volunteers on an evening walk out to where the lion enclosures come together in a sort of intersection. An alley runs between the enclosures, and within the four enclosures, at least eight lions reside. At sunset, Frikkie leads the group to the space between the fences and asks them to sit—no whispering, no noise at all. Soon, the lions come to the fences, interested in this gathering. Along with the volunteers, the lions see each other. As the sun sets, they begin to communicate—slowly at first, with a coughlike sound that lets the other lions know they are there. Another lion responds with more of a growl, and the responding voices escalate to roars that cause the volunteers' eyes to open wide, and their hands to reach for their hearts—where they can feel the sounds vibrate in their chests. The roaring competition continues for about ten minutes—both the males and females claiming their territory just five feet away from the humans frozen in awe. Then the voices gradually subside to a heaving, a gruff breathing, and finally silence.

Volunteers are never so quiet as after one of these performances. Most choose to walk to their cabins in solitude, considering what they have just experienced, contemplating their own fragility in the world, and knowing nothing in their lives back home can compete with dueling lions. If they speak, it is in a whisper, almost as if they have just been to church, sharing a hallowed moment.

Not all bonding experiences are so sacred. Frikkie knows young people, and he knows they need diversion after they work hard. Sunday afternoons are always reserved for a game of soccer with the Bushmen. The Europeans, who have been playing soccer since they could walk, show up in their expensive shoes, outfits, and wristbands while the Bushmen show up barefooted, sometimes shirtless, and already dirty.

The rules are flexible—sometimes the players have to play around a zebra named Zibi or a donkey named Jasper, the two ostriches that come to see what the fuss is about, or even a warthog or two, snorting through the crowd, with their sharp tusks clearing the way as they prance through with their tails held high. Distractions like this always hold surprises, but one outcome never changes—the Bushmen always win. Sometimes they'll lower their playing skills enough to make the volunteers think they might have a chance—to have some confidence in their abilities—but the Bushmen always conquer in the end. It is a humbling culture lesson that stays with the European volunteers.

Frikkie tries to pass on his own experiences in the bush to the volunteers, teaching them survival techniques and directional skills. One of his favorite activities is the bushwalk and sleep out with the volunteers. Sometimes he takes groups on those walks—very fast walks. Most volunteers will attest to the fact that this man in his 60s can outwalk any of them—through the bush, pointing out plants and roots that could be eaten, animals' tracks and spoor, and moss on trees that marks direction. Sometimes, he'll make them find their own way back to the campsite or to Harnas or some other specified spot. Without their GPS systems, most people can't find their own homes these days, and often this chore is one of the hardest for the volunteers because of their lack of experience in mere observation. In April 2006, one volunteer filed this diary entry:

> Frikkie asked us to find the way back to the campsite on our own! Well, in Europe this might be only a matter of minutes. But in the African bush land, every bush looks alike and therefore finding your way back can be a real problem! We barely remembered if we were walking towards or away from the sun on our outward journey—the sun rose in various places that day! Making 20 people agree on which direction we should take also is hard work, as everyone thinks they know best! We finally reached some sort of agreement (with a few stragglers trailing behind us), and then we realized that we had no water and wouldn't be able to walk around for a long time. But we made it! And we had taken a direct route, which reinforced group morale.

While out in the bush on a campout, Frikkie uses the silence, isolation, and darkness to teach these young people to look around them and appreciate what nature has to offer. Although most of the time, the group sits around the campfire, at some point Frikkie asks them to get up and follow him into the darkness of the bush night. One online entry describes this "heavenly" experience: "As the darkness came, the 'African sky' appeared. Frikkie asked us to turn off all our lights, and he took us on a little walk away from the fire. We lay down and everything was completely silent. We could see the Milky Way, Scorpius, and Jupiter. It really was one of those days where you think you could stay here forever." Another volunteer, Juli, from Switzerland, tells her own tale of such a night: "Frikkie stopped sometimes and philosophized about the stars and life as such. An indescribable feeling was arising inside me, as I was sitting there, small as I am, in total silence under the eternity of the African sky, making me feel small."

Frikkie is sensitive to the needs of each group, which he knows has its own personality. When he senses the volunteers have been working especially hard, and he knows they need not just a bonding experience but also an outlet to their frustrations, he organizes a special game similar to the American television show from a few years back, *Fear Factor*. Because the volunteers face personal fears every day—walking into the wild dog enclosure, feeding a cheetah, putting a pill down a resistant monkey's throat—the game is perfect to test their newfound confidence. Those who have been at Harnas the longest help organize and run the game, and Frikkie stands by, making sure no one is ever in real danger. Volunteers are divided into two teams, and the contests begin.

The challenges vary according to the situations available. They sometimes include eating a big plate of baboon food and drinking warm goat milk. Members may have to walk into the teenage baboon enclosure as a group, retrieve a can of soda, and get out—all without getting pounced on by the playful inhabitants. If lions are available and still young enough, volunteers might be timed on how long they take to pull a ribbon out of the tree above the platform where the drowsy lions are lounging. Perhaps the group will have to retrieve another ribbon out of the wild dog enclosure, using what knowledge they have gathered about avoiding attack by staying together as a group. Even the crocodile enclosure can be in play if the giant reptiles are sleepy enough. The games are all in fun, and no one is forced to do anything he or she feels too uncomfortable with—although team points may be subtracted for nonparticipation. The winning team gets a formal breakfast in the lapa instead of the self-made one they eat at the Volunteer Village.

Frikkie understands the kinds of experiences that unite a group, and he uses this motivation effectively. For many of the volunteers, he becomes a stern father figure whose respect they work for. As Mikkel knows, "If you do something that's a little scary or hard, you'll grow in confidence. Once Frikkie made me go into the wild dog enclosure and give them pills. I was scared, but it was awesome. And if Frikkie compliments you on it, you know you're really doing something good because he doesn't give out compliments very often. When he does, it means something." Another volunteer, Mitra, calls him a "surrogate father, uncle, mentor, and friend to each and everyone who passes through Harnas. His words of wisdom and stories allow you to think about who and what you are, where you've been and what direction you want to take in your life, to think about what's important in life and where our priorities should be."

The wild animals are the biggest fear to overcome for many people, but not always—and Frikkie knows this, too. Consider, for example, the story of Marcel from Germany, who chose Harnas for the reason most volunteers do—to work closely with wild animals. Marcel is a quiet, shy young man who picked Namibia because of its German background. He figured there would be more German speakers in Namibia, and he wouldn't be required to speak much English. Because most of the volunteers speak English, however, he was wrong about that. Marcel arrived at Harnas in November 2007 and stayed for four weeks. Nothing special stood out about him during this time. As in the rest of his life, he tried to do his job and stay out of the limelight. But just two days after returning home, he knew he had to go back to Harnas. He was not finished with it—and it wasn't finished with him. He says he missed everything: nature, freedom, animals.

When he returned to Harnas just a few months later, Marcel faced different challenges and experiences than he had during his first stay. This time Frikkie noticed in him traits he needed in volunteers. In return, Marcel watched Frikkie closely and tried to learn from him. He worked hard and enjoyed it, and little by little Frikkie came to depend on Marcel as his right-hand man. Marcel stayed for three months, and with special permission, got a work visa to stay three more. When Frikkie had to leave for a week or so, Marcel took over running the volunteer program, a responsibility he never thought he could undertake.

When Marcel was asked about his greatest fear at Harnas, he didn't say "working with the wild dogs" or "cleaning out the lion's enclosure" or even "watching out for poisonous snakes." He said, "Standing up in front of 40 people and talking. That was the biggest challenge, not working with wild animals. People are scarier than any animal." When he goes home, he plans on getting a degree in animal biology so he can work all day with animals, and when he does so, he'll be a much more confident man than he would have been—both with animals and with people—had Frikkie not picked him out of the crowd to lead.

Marcel admits, "I've noticed that my self-confidence has increased. Just standing in front of 40 people and telling them what to do is new and hard for me. I know my strengths and weaknesses more. I learned how I am best and how other people see me. I know myself better. There's a moment when you have to be the bad guy with volunteers, and it's not nice for them or for me. You know that you don't want to be the bad guy, but you have to—just like with the animals. You have to show them their limits and keep their respect. You want them to love you, but you can't let them take advantage. You have to take authority and show them you are in charge."

With whatever time volunteers have left over after working and playing hard all day, they are encouraged to spend time with their animals. That is their constant assignment because Frikkie knows that the bond between one human and one animal is the strongest of all. On a peaceful weekend afternoon, when such work as digging a water hole or gathering old bones in a leopard enclosure is rarely planned, volunteers have the opportunity to go and simply interact with wildlife. Inside a monkey's enclosure, someone might read a novel while the creature perches on his shoulder or sleeps peacefully in a homemade hammock nearby. You might see a young woman writing in her journal with her head propped up against a cheetah's back while the animal sleeps in the afternoon heat. Impromptu soccer games between lion and man have been known to occur, and baby baboons are always ready for a walk in the bush with a group of human friends.

Marieta—who, if there is such a thing as a baboon-whisperer, certainly is one—loves to take her baboons for walks. Since over 80 percent of the baboons at Harnas have been hand-raised by Marieta herself, they will follow her anywhere, like obedient children. Volunteers are eager to be invited to attend one of Marieta's baboon walks, and Kat from the United Kingdom describes with delight her own experience:

After a relatively normal day at Harnas of feeding and cleaning our animals, and patrolling the enclosures, it was time . . . for a baboon walk with the teenage baboons. With half of the group on game count, there were about 17 of us standing apprehensively at the bridge as Marieta let the baboons out with instructions to us not to scream or run, and to freeze if they bit us. Those with experience walking the baby baboons (with scars to prove it) were waiting to see how things could be with bigger versions. However, the baboons ran past us relatively uninterested and seemingly excited at their freedom— but never straying too far from the group.

They did manage to topple a trash bin, only to find a Coke can, with the proud baboon attempting to drink from it. Otherwise they followed or led the way down the airstrip. Periodically a baboon would come over with his hands out to be picked up, and they would sit on your back, head, or, like a baby, rest in your arms until they wanted to be put down again. Amongst themselves there were squeals and fights, but except for some "poo" we remained untouched, to all of our surprise. They were out for about an hour, definitely an enjoyable one! The most amusing thing was probably their fear of a plastic toy snake— from which they all jumped back a mile.

Marieta carries this fake snake with her in case one of the baboons gets out of hand. The ploy always works.

On a typical Thursday morning the transport truck arrives to take volunteers to the airport or to Windhoek. Suitcases and backpacks are loaded up in silence, and then the hugging begins. Some of the volunteers have been at Harnas for two weeks and others for many months, but no matter how long they've stayed, virtually all the volunteers cry when they leave, very often even the men. They exchange email addresses and vow to stay in touch. They know that several websites keep volunteers connected after they leave, and many of them will take advantage of that resource.

They have already said good-bye to their favorite animals—a much more private leave-taking. They have trained and passed on the responsibilities of their animals to volunteers who will stay on, but that doesn't mean they stop worrying about them. More tears are shed when the volunteers are saying private good-byes to leopards and cheetahs, baboons and monkeys than among people at the transport van. After they go home, the stories they tell to their friends and families don't single out the friend they met from the Netherlands or the cool guy from Denmark who made them laugh. Their stories and their pictures will show the animals, and the dreams they have will be about the cheetahs and giraffes and lions. This isn't to say that the bonds they form with other volunteers are unimportant. They're just secondary to the real reason they came and the real sadness they feel in leaving.

The driver of the transport van is a patient man. He does this weekly, and he waits a long time for the hugging and crying to subside. Finally, though, he announces, "Okay, into the van!" and those leaving find their seats and settle down. During the seven miles to the gate, the crying continues, but as the van approaches the main exit, silence sets in as each volunteer contemplates driving through the gate, probably for the last time. At this moment they realize that nothing in their lives will ever touch this experience. Harnas sets a mark that will be hard to beat. The volunteers know they are different people than when they entered the gate for the first time.

The lessons they've learned are varied. Sally remarks, "You find this part of yourself—with your outside appearance being less important. At home I couldn't leave the house without looking perfect, but here animals don't judge you for your appearance. You just become your real self. You lose the other, fake part. You don't worry if you're pretty enough. . . . I want to stay this person. It's so much better than the person I was before." Mikkel, on a different note, goes home with a broader perspective of the world and his comfortable part in it: "I learned about hard work. And I learned that Africans don't eat like we do and don't have what we have—and yet they're still happy. Even when they're working so hard, they laugh and have fun with life." One young woman claims that now she "is a leader," and a young man realizes that he is "capable of much more than I ever anticipated."

The list of positive changes goes on and on, and while one group is leaving, sniffling as they look out the window for a final glance back, more volunteers are flying in from all points of the compass, excited, nervous, and unsure about what to expect and what will be expected of them. Calmly smoking another cigarette, Frikkie waits for them, curious to see what problems and insecurities this next group will offer, ready to break them down and build them up, but knowing that his greatest partners in the process are the animals, who also calmly wait for the next caretaker to love them.

Nico preparing for a bike race

chapter thirteen

SON, BROTHER, HUSBAND, FATHER

*Do not measure your loss by itself, if you do it will seem intolerable;
but if you will take all human affairs into account you will find
that some comfort is to be derived from them.*

—Saint Basil

ALL THREE VAN DER MERWE CHILDREN GREW UP STRONG
and independent, and they showed compassion and care for
those around them, whether animal or human. Besides offering
their time and energy on Harnas, the children and their spouses
often took time out to work for other causes as well, an extension
of their love for Namibia and all of its inhabitants.

Nico loved to participate in charity events, especially for children who grew up without the advantages he had enjoyed. Athletic events were somewhat limited for him, however, for two
reasons. First, when he was six years old, he contracted rheumatic
fever, and doctors cautioned him against overexerting himself
because of subsequent damage that might have been done to his
heart. He took heart medicine until he was 21. At that point

doctors performed tests and found his heart to be sound, so he was taken off the medication.

The damage to his hands from the plane accident also limited him to some degree. He couldn't play some sports—rugby, for example. Cycling, though, was easy for him and he loved it. He often cycled distances of more than 60 miles just for the enjoyment of it, as well as training for races.

Because his job at Wilderness Safaris was also very physical, not surprisingly, when he visited Harnas he used this time to relax and sleep. Marieta remembers that sometimes he would show up and sleep through the whole next day. Even though he seemed fit, he was still out of breath a great deal, but everyone knew how hard he worked and assumed it was the norm for Nico. His body was holding a secret, however, and in retrospect everyone wondered why they didn't see the clues.

In December 2006 on a hot, sunny summer day, Nico and three female co-workers participated in a charity bike race for orphaned children. Many companies from all over southern Africa were involved, and Wilderness Safaris was one of them. The route was from Windhoek (inland) to Swakopmund (on the ocean), a total of 350 kilometers (about 220 miles) through mountains and desert, mainly on gravel roads. The course wasn't easy and the temperature was expected to soar over 100 degrees, an added burden that would push all of the racers to their limit. The race started at noon, and Nico decided to cycle the first 60 or so miles to make it easier on his female teammates, who could then cycle when it got cooler. At the 20-mile mark, Melanie, with daughters Morgan and Nica, met Nico, gave him a banana to eat, reapplied sunscreen for him, and sent him on his way with kisses.

Thirty miles later, the route reached a plateau with a spectacular view—mountains and desert around and behind him and the ocean in front of him. Nico took a deep breath, absorbed the beauty, and smiled, knowing the rest of the way was generally downhill. The road was still rough and difficult, but he pedaled on. The heat kept rising to over 110 degrees. On one side of him, the road dropped off in a steep cliff. On the other side was African bush filled with thorny vegetation. The combination of the two dangerous shoulders kept bikers pedaling straight ahead, and Nico concentrated on the next hill, the next turn, the final downhill coast.

Suddenly, Nico felt faint. His chest tightened and he found himself trying to catch his breath while he fought the pain that was increasing as his chest constricted. He lost control of his bike and drove straight into the bushes. Trucks and vans, some full of bikers and others with supplies and water were sharing the road, and one woman in a car saw Nico fall. She yelled at her husband to stop. They went back and found Nico struggling to his feet, trying to pull his bike out of the bushes, covered with scratches and punctures from the thorns and burrs. He was completely unaware that he had just suffered a heart attack, believing only that he was dehydrated and dizzy from the heat. A man his age doesn't consider a heart attack when he feels ill. Heart attacks are for older people, not for well-trained athletes in their 30s like Nico.

"Are you all right?" they asked. "Can we help you?"

"I feel like I'm going to faint," Nico replied. "It's so hot. I can't breathe, and I think I might be cramping."

They helped him to the front of their car, sat him sideways with his feet outside, turned on the air-conditioning, and gave him water. Fortunately, they had a cell phone and called for help. While the woman was calling, without a sound of warning, Nico slipped off the seat of the car and onto the road. He had just suffered heart attack number two. The couple jumped into action, lifting him back into the car, and calling again, this time for an emergency truck. They began to drive as fast as they could toward the approaching truck, but the road was difficult, and they couldn't make more headway, especially since it was also full of bikers they had to avoid. The emergency truck met them, but it was not equipped for this kind of emergency. They put Nico in the truck, gave him water, and tried to address his symptoms because they weren't sure what the problem was. Then it happened again. Nico's body contracted with pain as he suffered heart attack number three. They stopped the truck, took him out, and laid him on the hot road while they tried to revive him. Nico's eyes fluttered, he seemed to wake, and they put him back into the truck and sped back toward Windhoek.

An ambulance from the hospital in Windhoek was also on its way, driving as fast as the road would permit, and when the two vehicles met, the paramedics transferred Nico to the ambulance. They checked him over quickly, suspected a heart attack, and gave him a shot of adrenaline. They did not have a defibrillator. The ambulance began the drive toward the hospital, but again, they couldn't go very fast.

While all this was happening, someone contacted Melanie, and the first person she called was Marieta, back on Harnas.

"Ma! Ma!" she cried. "Nico has had a heart attack and is going to the Windhoek hospital! You must come!"

Jo was out on the far edge of the property conducting the tour for guests. Marieta radioed her and told her to end the tour and come back to the house immediately. Jo raced back and suggested they call Rudie, Marlice's husband, because he is a doctor and could help them get information. Marieta called him on his cell phone and relayed what the ambulance workers had said about Nico's having had a heart attack.

"Don't worry," Rudie said, trying to calm them. "It's probably just dehydration. They'll just put him on a saline drip and he'll recover fine."

Marieta hung up, feeling much relieved. Dehydration was a problem everyone living in Namibia understood, and she knew that Nico could survive it and be all right in a matter of hours. She and Jo hugged and tried to relax. Dehydration—not a heart attack. They tried to breathe normally again.

Nico was a fighter. He had inherited the stubbornness and psychological strength of his parents. But he wasn't just dehydrated. His heart, damaged so long ago by rheumatic fever, was giving out under the immense strain of the heat and exertion. No one was aware of it at the time, but his heart had grown twice as big as it was supposed to be—fighting to pump the blood through his body but unable to do so. At the hospital, Nico continued to fight through yet another heart attack, but the doctors couldn't save him. After four heart attacks in three hours, Nico died. He was only 35 years old.

At Harnas, the radio beeped. Juan, the guest manager, was calling from the lapa office for Marieta. When she picked up the radio, Juan said just one word: "Telephone." And then she knew. Marieta remembers, "I can't tell anyone what it feels like. Losing a husband is a huge thing, but losing a child! You're not supposed to outlive your children. We looked back and remembered how tired he always was. But that's all in retrospect. If he'd gone to the doctor, they might have caught it, but this is the way life is."

The week before the race, Marlice and Rudie had spent four days with Nico on the coast. Looking back, Nico's sister feels this time was "a blessing" because he was living on "borrowed time." The two of them had talked for hours and had taken relaxing walks on the beach. When Marlice and Rudie dropped him back in Windhoek, Nico had casually mentioned that he was riding in a charity race the next day.

"Are you fit?" Marlice asked.

"Probably not," Nico replied with a shrug of his shoulders. He went on to say that he hoped to ride the final leg of the race, the easiest, downhill into Swakopmund. The payoff for the race would be staying in a luxury hotel that night, free to all participants.

"I'll probably have a heart attack," he added with a laugh, "but the freebie hotel is too hard to pass up."

Nico's death had a profound effect on Melanie. She hadn't always seen eye to eye with other members of Nico's family, and she tended to show her temper when she felt Nico or their daughters were shown any disrespect. As the quietest sibling, Nico had been much like his father—the peacemaker who would rather concede than continue an argument. Melanie was more like Marieta, bold and quick to make right any injustice to her or to her family. Her similarity to Marieta was both an asset—as Nico knew how to live with fiery women—and a disadvantage—because her temper and bold tongue sometimes led to conflict.

The rest of the family admits they were concerned about Melanie after Nico's death. She suffered severe depression and was put on powerful drugs to help her cope. So when she phoned Marieta and said, "Nico must come home to our house before we bury him. He must sleep one more night in the house he built for me and our children," Marieta was too intimidated to say no. She thought, "Oh my God! I can't do this!" but she was afraid that if she said no to Melanie, it would cause a rift in the family that could never be healed. Rather than crying and grieving, Melanie acted unnaturally calm, which also made the rest of the family uneasy. Melanie believes her resilience stemmed from more than the medication the doctors had prescribed to help her through the trauma of losing her husband. She knew she had developed an inner strength and peace because she *had to* for Nico. Marieta didn't know how to react to this sedate Melanie, so she just agreed. Fortunately, it was the right decision.

The day of the funeral, Melanie called Marieta again. "I need you to meet me at the funeral home to prepare Nico's body," Melanie said. "I don't want some indifferent undertaker to be the last person to care for Nico."

Again, because of Melanie's calm demeanor, Marieta didn't know how to react, but not knowing how to turn her down, she went.

"At the time," Marieta recalls, "I had a very small baby baboon I was nurturing, and when a baboon is so young, she must go everywhere with her 'mother'—me. Usually, since I spend most of my time at Harnas, this isn't a problem, but taking a baby baboon to a funeral home and then a funeral was a little unusual—even for me."

Marieta imagines that people assumed Nico's death had affected her brain. "I figured that people looked at me and thought, 'This woman is crazy! She brings a baboon to her son's funeral!' And while Melanie and I were preparing Nico's body, the baby baboon sat on my feet the whole time in the room.

While we washed him and dressed him, I cried my heart out. I cried until I felt I couldn't cry anymore."

For Melanie, preparing him for his funeral was a spiritual and loving act for the two most important women in Nico's life: the mother who had brought him into this world and his wife and partner on the last part of his journey. Melanie insisted the funeral home supply them with warm water to wash Nico because he had hated to be cold. She used his favorite soap and brand-new soft towels to dry him. She dressed him in the clothes he was married in, and with Marieta's help, she washed Nico's hair and gave him a manicure and pedicure because, she says, he used to love that kind of pampering.

A few days later Melanie put up a white tent at her home, lit candles, and put out flowers. "She had food and drink enough for hundreds of people," Marieta remembers, "for a celebration of life rather than a wake."

Nico's coffin had been made by his friends from work because he would have disdained a fancy one from the funeral home. They placed the open coffin right in the middle of the living room where everyone could see this young, beautiful man who had died too soon. All of Nico's favorite things were nearby: fish swimming in containers, whiskey, candy, his bicycle. All evening long, one or another of his friends pulled up a stool to the coffin and spoke to him, saying a farewell. If someone wanted to make an observation about Nico, he or she rang a bell to call everyone's attention—the old dinner bell that had originally hung on his grandfather's farm. Throughout the evening, people told funny, sad, wonderful stories about their adventures and journeys with Nico. They told about quiet hours with him. They expressed their love.

They drank and laughed and cried together. Into the coffin, Melanie put some whiskey, some candy, and a letter to Nico's father, asking him to look after her husband. And the people of the Nama tribe, who had worked with him, sang and played drums all night outside the house.

Marieta admits that when she first heard what Melanie wanted to do, she resisted. "I thought, NO! But afterwards, most of the people told me that it was the most beautiful funeral they had ever seen, and they wanted to do the same things for their families. Also, in the end, I came to believe that if I hadn't gone through all that, I couldn't have said good-bye myself. But because I held his face in my hands and had the whole night to say good-bye, it helped. It's not *all right*. It can never be all right again, but it was *better*."

Nico had loved classic cars, so the next day his coffin was carried in the back of a 1956 Chevy truck. Melanie laughs as she remembers her daughter Nica. Nico had bought a SpiderMan outfit for Nica for Christmas, so the three-year-old decided to wear it to the funeral. What an unusual funeral it was—one daughter in a SpiderMan costume and the mother of the deceased carrying a baby baboon. When they took him to the crematorium, true to her vow to see Nico through to the end, Melanie pushed his body into the fire.

Later on, the family put up a cross at the spot where Nico had his first heart attack and fell. In true Harnas fashion, Marieta had with her a new, different baby baboon she had just rescued, but the baboon had been raised by a black family and wasn't yet comfortable with Marieta's white skin. So Melanie had brought her black gardener, Stetson, to hold the baby baboon and even sleep with him the night before. Photos from that day show Stetson with a baby baboon wrapped around his neck.

In honor of her husband and her son, Marieta had a small open-air stucco chapel built on Harnas, away from the house, out toward the Volunteer Village. Stone pillars hold up a corrugated tin roof. The structure is open to the African breezes, and a candelabrum fitted with real candles hangs from the roof. Log benches line the center space, and at the front of the chapel is a stone altar with a simple cross adorning the front wall. At the back are two carved wooden crosses hanging on the wall, each with a photo on a long shelf below. Candles stand on the shelf as well, with small porcelain angels and a guestbook. Walking into the chapel, visitors feel a refreshing coolness in contrast to the hot African sun. This, like the rest of Harnas, is a place of peace and healing, a place people can come to in groups or individually to sit in silence, to consider their lives and their choices, or just to enjoy the sense of stillness and compassion in honor of two men, father and son.

The loss of Nico is one more tragedy this family has had to endure, and yet Marieta tries to find reason in it. "His death changed Melanie's whole life," she explains. "Now she is so kind. Melanie and I are such good friends and she says that we are her family. She is like a daughter—as is Jo. My sons' wives are so good."

Schalk has also felt his brother's death intensely. The middle child and the second son, Schalk was allowed to run free and be the wild man he comes to naturally, but with the death of first his father and then his older brother, Schalk is now the man in the family. The death of his brother made him keenly aware of his own mortality, and since Nico lost the opportunity for himself, Schalk wants to make sure he gets to know his own children well.

Nica, who had spent most of her first three years with her father—first as an infant in a backpack, as he trekked through the wilderness as part of his job, and then on foot, following him everywhere—was traumatized by Nico's death. Nico had been the primary caregiver for his daughter, whom he adored and had fit so easily into his daily schedule. Melanie had been available, too, but Nico and Nica had a special bond. When suddenly her father was absent from her life, young Nica felt abandoned. Even at three and half years old, she knew that her life had been forced to take a different path. Instead of turning to her mother, she turned to her *ouma,* and she asked to go live at Harnas. Melanie agreed, and since her older daughter, Morgan, was in school in Windhoek, Melanie stayed there and continued to work.

Nica bonded deeply with her grandmother, sleeping with her, following her around during the day when she wasn't in school, and at first choosing not to interact with people outside her family. But Nica has gradually begun to thrive at Harnas, learning to love the animals just as the rest of her family does. She plays with her two cousins, Samar and Aviel, and she is surrounded by people who love her and animals who adore her. But Melanie knew that breaking up the family in this way was not a good idea, especially as she and Morgan needed to be a closer part of the extended family, she sought the peace and healing that Harnas offers.

So in spring 2008, Melanie took a position at Harnas, running the guest relations program, which was foundering a bit. In fact, Marieta had temporarily closed the guesthouses because taking care of the tourists was interfering with the main mission of caring for the animals. As Frikkie has with the volunteer program, Melanie has made the guest program efficient and profitable. Already she has shown herself to be a vital part of the Harnas family.

Melanie is thoroughly experienced in working with guest lodges and with workers, and her presence quickly made a difference as she interacted daily with the servers and cleaning staff. In addition, Melanie has used Harnas as a place to train future employees for a new government lodge to be built inside the Etosha National Park. Melanie interviewed over 300 people in Windhoek and chose 40 to bring to Harnas to train for the opening of the lodge. These trainees come from all culture groups inside Namibia—and even include two Zimbabwean refugees. Melanie trained them in lodge service, restaurant cooking and serving, tour guiding, and so on. While Melanie claims she could have trained these people anywhere, she chose Harnas because "it has a soul" and many of these people need to find "their own souls." She also believes that being around nature brings out the best in the workers because they experience the "true Harnas spirit" that only the natural world can offer. Many of them have previously regarded wild animals as either a nuisance or a food supply. In their new jobs at Etosha, they will need to protect the many wild animals in the preserve.

Melanie's presence has also had a positive effect on Nica, who has bonded with her mother now rather than just her grandmother. Morgan visits on the weekends and during school breaks, and both daughters have become true horsewomen, following the example of their uncle and grandmother. Their determination to ride alone is reminiscent of the childhood aims of Schalk and Marieta: "No fear—and if you fall off, you get right back on." Nica pouts if someone wants to ride with her or hold the reins. She prefers to ride alone, and to gallop at full speed, bareback. Morgan already has her own horse and rides like a pro. Stubbornness, skill, and determination to succeed have certainly been instilled in the newest generation of van der Merwes. With the enduring family ties to Harnas, what Marieta has created will last a long time after her generation is gone.

A large picture of Nico on his bike hangs on Marieta's living room wall, and at times Melanie is overcome by having lost the man she calls her "soul mate and best friend." On anniversaries of his death, his birth, their marriage, and so on, Melanie still has a hard time getting out of bed and functioning, but little by little, with Marieta's help, she is healing. Marieta is an apt guide because she understands loss—first her mother, then her father, her stepmother, her husband, her son, and many friends, including, of course, so many of the animals she has loved.

After less than a year at Harnas, Melanie decided to split her time between Windhoek and the sanctuary to be able to spend more time near Morgan. Nica is happily ensconced in Harnas and will stay with her grandmother for now. As is often the case, Harnas is a place that people—both family and visitors—move in and out of as their needs direct them.

Marieta's final words on the subject of Nico's death distills her whole philosophy of Harnas and life in general. "Sometimes people say I'm such a survivor, but it's not me. I turn to the animals for healing. Like Father's Day, Melanie cried all day and asked why I didn't cry. I said, 'No, Melanie. I don't cry. I cry in my heart, but no amount of crying will bring them back.' I'm actually very lucky that I'm here with these animals. Among millions of people, I was God's choice to live with and look after these animals. Then I can't say I'm unhappy or that life isn't good. It's good! I believe in it."

Jo and Mara discuss a lesson plan for Cheeky Cheetah

chapter fourteen

COMPASSION *and* COMMUNITY

May all that have life be delivered from suffering

—Buddha

MOST MORNINGS, IN THE CHAOS OF DAILY PREPARA-
tions, Marieta's leader of her Bushman workers will poke his
head through Marieta's door, say something quickly to her in a
low Afrikaans voice, she will answer, and he will disappear to follow her instructions. Notoriously shy, Bush people slide around
Harnas quietly, sometimes doing what they're told, sometimes
not. They are still getting used to the idea of "work," a concept
that never entered their traditional culture before about 1960.
Life was just life: you did what you needed to do to survive.
Doing a specific job for money was simply not a part of their
thinking. Money means an extra step between activity and food,
something they never had to worry about before. Even though
they have good jobs at Harnas, many of them still leave, take

another similar job somewhere else, and then move again, often circling back to Harnas and then repeating the cycle. They are adapting a nomadic existence in a country that hasn't much free space left to roam without running into fences and private farms.

Bush people rarely talk to anyone except the family and other Bushmen, and they seem as wary of others as others are of them. If a volunteer or a visitor does engage them, however, they respond with wide, beautiful smiles and a friendly nod or "Hello." It seems they are waiting for others to take the first step, and then they are pleased to be noticed. Most of the Bush people at Harnas are also called *San*—meaning simply "the people"— and are believed to be the very first inhabitants of this land, long before history recorded them. With European civilization, division, and conflict that came as a result of colonialism, the San lost their traditional way of survival and have not, as yet, figured out how to live on their own terms in this new world.

They still struggle with the idea of showing up for work every day, but Marieta has a core group of Bushmen who have been with her for many years. As new workers come and go, she can always count on her leaders to train them. It is a constant struggle, though, and Marieta often mutters under her breath about the inconvenience of having to check up on them constantly. But to make Harnas work, she knows that she must involve all aspects of the community.

Although the sanctuary was created with the idea of saving the wildlife of Namibia, Marieta and her family realize that saving animals is just a part of saving Namibia as a country and a culture. Early conservation movements in the 1960s and 1970s often failed to include the human community in their efforts. As a consequence those early projects were doomed. Marieta and

Harnas—and other movements like it—knew they must include people in the vast web of improvements to the area, improvements such as education and health care. Once native people see the benefits of a holistic approach to conservation, they are much more willing to include wildlife in their concerns. Once they themselves are not hungry or sick, they can participate in the general renovation of an area and take part in what has become the new lifeblood of Namibia: tourism.

Although this development in a country wracked by poverty and AIDS might seem an insensitive endeavor, the income derived from tourism fuels employment, improves education, and generates money for health care for the local people. Foreign and domestic money contributes to the exchange of cultural advances and awareness for all involved. Tourism is now the third largest source of foreign exchange, after mining and fisheries, and it is expected only to grow. Because tourism in a place like Namibia depends hugely on the health of the environment and its wildlife, local people will begin to see that their welfare and the future of their children depend on protecting the animals of their country.

Marieta's most vital project for the local people began at Harnas on a cold morning in August 2003. The family was awakened by a knock on the back door of Marieta's house. A Bushwoman was holding her malnourished, unresponsive baby, who was having difficulty breathing. Fortunately, Marlice was visiting Harnas with Rudie, who immediately phoned for an ambulance from the hospital in Gobabis. This happened during a holiday weekend in Namibia, and the driver was less than eager to drive out to Harnas to retrieve the sick child, especially once he heard the baby was San. His reaction was a familiar one. In the hierarchy of race in Namibia, the San people barely register. So after trying everything medically they could do at the farm, Rudie drove the child to the hospital himself.

As they arrived, the child went into cardiac arrest, and Marlice ran to the nurses station, where three Herero nurses were seated. They listened to Marlice describe the situation but didn't offer to help in any way. They didn't even get up from their chairs once they understood the child's race. "After all," their behavior implied, "it's just a Bushman's child." With the medical supplies and equipment available, Rudie worked to revive the child himself but was unable to do so, and the child died.

Marlice and Rudie were outraged. She had grown up with Bushmen children and felt closer to them than to her white friends, and Rudie had spent his career as a doctor trying to help the suffering populations of nonwhite Namibians. Everyone knew that the health care needs of the San were neglected and ignored (as well as all their other needs), but this incident spurred Marlice and her whole family to do something about it. Rudie and Chris Heunis, a pharmacist, decided to create a permanent health facility within reach of the San population, and since that population lived all around Harnas, the area near the farm seemed the perfect location. They identified a building within a few miles of Harnas that had once housed a clinic, and then they began to search for, beg for, and collect donations to reopen it.

They also conducted studies to identify the major health problems of the San—information that no one had ever gathered. The Bushmen and their families had been so ignored that their basic needs and problems were not even known. These people had lived off the land for so many centuries that it was almost impossible for them to think of living any other way. And yet to survive, they have had to learn new ways. The land they once inhabited was confiscated, sold, farmed, and fenced off. Like the Native Americans, they had no concept of *owning* land before it was taken from them. The game they hunted with bow and arrows were taken more easily with white men's rifles. And the survival skills they had perfected in balance with nature became irrelevant.

In her book *The Old Way*, Elizabeth Marshall Thomas, daughter of the anthropologist Laurence Marshall, who first studied the San of the Kalahari in the 1940s, laments that "Today, no [Bushman] lives in the Old Way," and in Namibian culture "Bushmen are the poorest ethnic group by far, with the greatest unemployment." But the social threats to the Bushman have been even more devastating than the economic ones. Not only have they been displaced from their land, but they have been the target of ethnic cleansing by both black and white populations, reducing their numbers from a high of 100,000 to just under 40,000 today.

Rudie's studies revealed that more than half the Bushmen are jobless because the idea of "work has no cultural meaning for them. They don't understand 'work' as a source of income as they seem to not understand the importance of money. Food is the only thing that seems to matter." Some of the women earn money for themselves by making and selling crafts and jewelry, and the San are a highly egalitarian society, so the fact that women earn more money than men has no real social consequence.

Alcoholism is endemic, with up to a third of their income spent on alcohol, which has a keen influence on the poor nutritional status of the Bushmen. They brew their own alcohol at times—producing a drink so strong that it "inflames the optic nerves and causes blindness," as well as liver problems. Alcohol also contributes to social problems like violence, promiscuity, rape, and the breakdown of the community. This problem is perpetuated by the Herero people, whose population outnumbers the San by eight to one. The Hereros employ but exploit the Bushmen, often paying them with alcohol for their work. The desire for alcohol is so strong now that when Bushmen are given free medicine for various illnesses, they sometimes sell that medicine to buy alcohol. In cities and towns, fathers send their children out to beg for money—sometimes even refusing gifts of food—and the money goes immediately to buy alcohol for the fathers.

Because many men don't work on a regular basis, and because of the high rate of alcoholism, San people live in abject poverty. Even the meaning of wealth is different in their culture. Bushmen believe children are wealth, and so birth control is ignored. They do not understand that having fewer children will give them a higher standard of living.

Finally, the disease that plagues all of Africa is especially widespread in the San population: HIV/AIDS rates are astronomical, but precise statistics are not even known because the communities are scattered and rural, and the San do not usually come to doctors for help. When doctors do see them, they often diagnose tuberculosis—indicative of immune systems that are compromised. And if HIV/AIDS is found, locating previous sex partners for detection and treatment is virtually impossible.

To all the van der Merwe family, not just Marlice and Rudie, these facts are heartbreaking, so with the help of donors, the Harnas Lifeline Clinic was born. On February 12, 2005, when Rudie opened the clinic, the staff included a doctor, a pharmacist, a physiotherapist, an interpreter, and an administrative coordinator. That day they treated 37 patients, five with AIDS. Word started to spread, and more and more patients arrived, both from the Herero and the Bushman populations. Rudie decided that a doctor would visit the clinic one Saturday a month, and the rest of the time, the clinic would be run by a registered nurse who could diagnose and treat many ailments. The first nurse, Anna Daries, had worked in rural areas for years and was familiar with the ailments suffered by the local population. She would pass on the more serious cases to the doctor.

On some Saturdays, the staff sees up to a hundred cases. Patients suffer from a variety of poverty-related diseases such as tuberculosis, malnutrition, lung ailments like pneumonia, high blood pressure, infections, and skin and eye diseases. As mentioned, HIV/AIDS tops the list, as it is now the major killer in sub-Saharan Africa. Close to 6,000 people die every day from the pandemic and 14,000 more are infected, according to the International AIDS Vaccine Initiative. Most of the at-risk population has no access to drugs, labs, or even preventive measures like condoms. In Namibia alone, almost 400,000 (19.7 percent) people live with HIV/AIDS, according to U.S. State Department statistics. In the Bushman population—where the disease is poorly understood and access to prevention lacking—the numbers are believed to be much higher.

When the doors of the clinic first opened, one of the first patients Rudie saw was a child named Anna. Her mother said that Anna was three years old, but she was extremely small and couldn't walk or speak. Knowing the poor nutritional state of most San families, as well as the adults' propensity for alcohol consumption, Rudie examined Anna's face carefully and saw several distinctive markers: her upper lip was thin and lacked the cupid's bow and the philtrum, the two vertical ridges that extend to the nose. In addition, her eyes were small and somewhat bulbous—with barely the crease in the eyelid that most non-Asian people have. The diagnosis was clear and heartbreaking because it had been completely preventable. Anna had been born with fetal alcohol syndrome.

Fortunately, Anna's condition could be improved—although never cured—with simple vitamins. She moved in with her grandparents, who were able to take better care of her, and they made sure she went back to the clinic every month for a new packet of vitamins. Although her IQ will probably always be low, and she may suffer heart, kidney, and bone disorders, her condition has improved dramatically from that first time her mother brought her to the clinic.

The battle for public health and native population stabilization cannot be won without people like Marlice and Rudie and the doctors and other volunteers who continue the fight. Eventually, they decided that the Herero people could afford to pay a stipend for their visits and for their medicine, but Bushmen would continue to receive theirs free. Some abuse of this system occurs—as it would in any country in the world—but the good the clinic does far outweighs the problems. Several local people have been employed by the clinic and a few have volunteered as well. Once the word got out about the success of the clinic, donations of food, medicine, and money started to come in.

The clinic's website offers these encouraging words:

This is a success story. . . . Since the clinic opened in February 2005, more than 1,000 patients have been treated. Diagnosis ranges from common backache, arthritis, and flu to more serious snake bites and epilepsy. Often, chronic diseases like tuberculosis, ischemic heart diseases, hypertension, and HIV/AIDS are seen. And the sight of two people was restored through cataract removal. Always people who don't have the money or transport to go to the nearest hospital for medical treatment are seen at the Lifeline Clinic.

Harnas is constantly changing, and the Lifeline Clinic has changed with it. The transition began later in 2005, when Marieta bought a parcel of land only 24 miles outside of Windhoek. The idea was to establish a second animal sanctuary where she could live out her "old age days" nearer to Windhoek with its hospitals and markets for fresh food for her and for her animals. Because Marlice's husband, Rudie, needed to live near Windhoek, where his medical work is centered, Marieta concluded that Schalk and Jo would take over the management of Harnas. At the time, Nico and Melanie were happily working for Wilderness Safaris and had no interest in the new project. Various other members of the family invested in the property, as did Marlice and Rudie's friend, Chris Heunis.

Rudie and Schalk oversaw the building of the new animal sanctuary, called N/a a'n ku se, meaning "God sees" or "God protects." As with other projects that have numerous investors, each with his or her own ideas of how things should be run, how money should be spent, and who should make the decisions, the building of N/a a'n ku se began to suffer from "too many cooks in the kitchen." Eventually, Marieta decided the tension was too much and the family fighting was making the project less joyful than she had hoped. At one especially tense board meeting, she announced, "This will not work. I think we should split the two farms up. I'll return to Harnas and live there the rest of my life." Shortly after, Schalk decided the same thing. He called his mother at Harnas and said, "I want to come back to Harnas. There are no men on Harnas, and I promised Dad while I was driving him to the hospital that I would take care of you."

So Schalk and his family moved permanently to Harnas, and Marlice and Rudie continued to live and work on N/a a'n ku se. In keeping with this new separation, the Lifeline Clinic was renamed N/a a'n ku se Clinic, although the same types and quality of care are offered.

Along with health care, the education of Bushmen's children has also been a concern of Marieta's family. Because so many Bushmen have lived and worked on Harnas—and have their families there—Jo van der Merwe wanted to start a school for the young children. She knows that the next generation of Bushmen children will need even more help to adapt to the fast-changing culture around them. The Bush families were not used to living in permanent homes, but traditionally had built temporary shelters made of sticks and grass wherever they were living at the time. To teach them about permanence and living in the way most Africans do, Marieta created a small village with homes, electricity, water and sewage facilities, and community space, placing it on Harnas property about a quarter of a mile from the family homes.

When San children apply to grade school in Namibia, they are usually turned away in favor of the Herero and other tribal children, who are much more prepared academically. Herero children are more disciplined, they have better health and hygiene, and they often have the preschooling that prepares them for the grade school curriculum. San children, on the other hand, have spent their childhood barefooted, running free. Most do not know any language other than San—and schools are taught in English and Afrikaans. They do not know their numbers or letters because their parents have seen no reason to teach them these things, never having learned them themselves. The cycle of ignorance and poverty continues in this way, and the schools—when they have a spot open—won't take the chance on a San child when there are better-prepared children waiting.

In response to this problem, Jo—with the blessing and help of Marieta—opened the Cheeky Cheetah Day Care Centre in 2004 with the goal of enabling farm workers' children to have access to some form of education, social development, hygiene, and health care. With proper training, San children can be prepared for the local grade schools when they reach the proper age, and Marieta and Harnas have paid for several children to attend those schools. At the moment, 15 Bushman children are being sponsored by Harnas and its supporters. The school fees, uniforms, and supplies are paid for through donations.

The first Cheeky Cheetah classroom was a small building erected between animal enclosures and filled with rickety chairs and tables; donated books, pens, and pencils; and, outside, a small playground. Jo herself taught the children, but their attention span was short, and the frustration level for both teacher and students was high. The project was worth pursuing, though, and Jo kept at it, fighting for donations and spending her time convincing San parents that it was a necessary step in their children's development. Students received baths on the lawn with a garden hose before class. They were given a good lunch to help them concentrate, and then they spent small bursts of time learning, interspersed with playtime outside. This happened a few days a week—when Jo could squeeze in the time in between all her other duties.

Progress was slow, but Jo's vision never wavered. Plans for a bigger, better-equipped building came to fruition in January 2008. A full-time, academically trained teacher was hired—Mara Kuhn—who had grown up nearby and therefore understood the ways of the Bush people. She speaks Afrikaans and English and a smattering of various other languages. The school now stays open five days a week, and the children have fallen into the routine of attending school the way students do all over the world.

"The school day begins with a good meal and then a warm shower in new inside facilities," Jo says, smiling. "The children are taught proper hygiene such as dental care and toilet manners. Their dry little bodies are moisturized with lotion and they are given fresh clothes to wear. Clean, fed, and content, the children begin school, and their attention span has improved beyond measure! They have books, pens, pencils, crayons, and even a television with educational videos—all from caring people who have made generous donations."

The struggle between the old and new is ongoing, and the cultural mores of the Bushman families often work against Jo's best efforts. One of the first Bush children to receive guidance from Cheeky Cheetah was the adopted son of Kasoepi, the woman who has worked for Marieta for years, and who helped raise the three van der Merwe children. Kasoepi had been unable to conceive but desperately wanted a child, so with Marieta's influence with the welfare agency, little Nanna, a one-year-old boy, came to live at Harnas. When he arrived, no one realized that he had a back problem that made it difficult for him to walk—although he persisted and eventually learned not only to walk but also to run.

It also became clear as he grew that he was not full-blooded Bushman, and together with the physical problem, Kasoepi struggled to accept him, and when she finally did conceive her own child, a beautiful and healthy baby girl, Nanna felt rejected again and so gravitated toward Marieta's family, and especially Marieta's grandson, Samar. The two boys became best buddies, and Nanna often stays at Marieta's house, even going on vacations with the family.

Nanna struggles, though, with having two families—one that sees the value of education and one that barely sees him at all, but Cheeky Cheetah has been of great benefit to him, especially since his best friend Samar goes to the school every day. Today Nanna is in second grade—a great achievement for many Bush children—but he continues to struggle. "I try to explain to him that he can change his life by his education here," sighs Jo, "but in his little mind he does not want to be in that school. I am not sure how to handle it yet; I don't know what to do, so for now I pray, try to convince him, beg him, and even threaten him in hopes that one little boy will get a better life."

Although precarious, the future of the next generation of Bush children—at least the ones who live at Harnas—looks much brighter with the persistence of people like the van der Merwes. Another teacher, this one for grade school children, has been added. Having an additional teacher to work with older children like Jo and Melanie's own children means that even if the Bush children cannot get into the local schools, they can continue their education at Harnas. In this way, Marieta and her family have extended their care of animals into the wider sphere where helping wildlife has led to other developments that contribute to a positive future for Namibia.

Caracals jumping for their breakfast

chapter fifteen

LIFELINES

*All I wanted to do now, was get back to Africa.
We had not left it yet, but when I would wake in the night,
I would lie, listening, homesick for it already.*

—Ernest Hemingway

THE ANIMALS, THOUGH, REMAIN THE MAIN FOCUS OF
Marieta and Harnas. One of the biggest dreams for the family is the
Lifeline Project, which has the goal of giving their predators—the
carnivores—"the life and freedom that they deserve but in a safe and
monitored area to ensure the long-term success of wildlife rehabilitation," as the Harnas website says. The goal of Harnas has never been
to create a zoo. In principle, Harnas "is against the confinement
of wild animals," but of the animals that are brought to Harnas,
only 25 percent can be released. They are too injured or sick—like
Audrey, the blind vervet monkey, or Sam, the lion with FIV. They
have been raised as pets and then given to Harnas and so have no
idea how to survive in the wild—like Afram, a cheetah whose owner
realized when the cub grew up that having a full-grown cheetah for

a pet wasn't really a good idea, or Jacob, the baby baboon that a woman owner finally realized she couldn't handle.

Sometimes they were orphaned as newborns and had to be hand-raised—like a family of meerkat babies found by a group of San children or a pack of bat-eared foxes whose mother was killed by a farmer who believed incorrectly that the foxes would kill his livestock. Other animals have been saved from slaughter—like Ina and Claus, the two crocodiles that were slated to be made into shoes, belts, and other accessories before Marieta brought them to Harnas. Yet those animals that *are* capable of surviving on their own do end up being released.

"My biggest wish," says Marieta, "is for my animals to be free!"

The Lifeline Project is designed to make that happen—to free as many animals as possible, but in an environment safe from hunters and poachers. That requires a space—a preserve—large enough (nearly 25,000 acres) for both predators and prey that is secure, with electrified security fences completely surrounding the gigantic area. The land must have natural and constant water holes and the right kinds of grass to support the right kind of prey animals. The large carnivores need to be fitted with GPS collars for monitoring, to make sure they are alive and well and receiving what they need to survive. Also, an antipoaching squad must be employed to make sure people don't hunt or capture Harnas animals.

Marieta's innovative but costly dream is now close to becoming a reality. She plans on releasing lions, a hyena, a leopard, wild dogs, cheetahs, and caracals as predators (and potential prey), as well as the springboks, hartebeests, and impalas, among other animals that serve as food for the carnivores. All these animals must be released in the proper balance and in the right order—prey animals first so they can get to know the perimeters and water holes. And then predators, a few at a time. What Marieta is trying to do is create a microcosm of what wild Africa used to be, and that is an enormous task. She is willing to put all her time, research, energy, and money into the project, however, because she really does want her animals to live as freely as possible in this new Africa that has led to the near extinction of many species.

Harnas is involved in other projects as well. The Wild Dog Project has been active since 1996, when several dogs were brought to Harnas for rehabilitation. One night a friend of Nick's from the police force called to say that four African wild dogs had been caught in traps up north on Ovamboland. The dogs had been chained to trees and left, without food or water, as bait for other wild dogs, which would be lured in by their pack in trouble. The farmer planned to shoot the dogs as they appeared.

Wild dogs are almost extinct in Namibia, so finding even injured ones was big news. Nick and Marieta didn't hesitate. Nick and the farm manager, Max, drove almost 250 miles into the bush and found the house in the middle of the night. Usually wild dogs fiercely resist being caught, and Nick didn't have a tranquilizer gun in those days. But these dogs were so weak—nearly dead—that he was able to throw sacks over their heads, release the chains, and put them into small cages in the back of the truck. He and Max drove immediately to the vet in Windhoek. All this occurred without incident—and also without Nick's telling the people who had trapped and tortured the poor creatures that he was taking their game. The two men spirited off all six trapped animals they found that night—four wild dogs, one hyena, and one African wild cat. All six animals survived, but every one of them lost the leg caught in the trap. They lived out their lives peacefully at Harnas, where the wild dogs served as the basis for the population that inhabits the sanctuary now.

The African wild dog, at times called the Cape hunting dog or the African painted wolf, has a lithe, long-legged body, standing about 30 inches at the shoulder and weighing 45-60 pounds. Large, stand-up, round ears help these animals sense vibrations and hear other members of the pack, communicating via the species' strange high-pitched squeak. One Harnas volunteer described the sound most aptly as like the sound made by rubbing a hand around on an inflated balloon. The dogs run in packs of 30 or more, and can wander for miles in a day, searching for food. The most striking characteristic of an African wild dog is its coat, which one guide-book describes as "like a tie-dye rug" of yellow, black, and white. Each dog has a unique pattern of the colorful mottled design, almost like a fingerprint, that enables researchers to identify individual animals.

Members of this species have been shot, trapped, exiled, and poisoned everywhere, even in the national parks, because of their reputation as uncontrollable and merciless killers, even though that is not completely true. The wild dog has been unable to adapt to an Africa where farming is more important than wildlife, and the population has dwindled as a result. The pack hunts during the day, chasing prey and switching out tired dogs for new ones until the prey falls in exhaustion. The hunt is extremely efficient and the communication between members of the pack is startlingly accurate, giving them one of the highest percentages of hunting success of all carnivores. And while they are relentless in their pursuit of food, within the pack they are unselfish. The packs are matriarchal, and litters of pups—sometimes up to ten in a litter—are raised cooperatively. The pups, plus injured and old members of the pack, are brought back food from a kill. Wild dogs do not abandon the weaker members of their pack because their strength lies in their power as a group, not as individuals—as opposed to leopards, for example, who live by the law of the jungle.

These animals are in real danger of extinction from the hunters and farmers who view them as vermin. Whereas they used to live in 39 countries, they are now found only in six. The only captive population of African wild dogs exists at Harnas, where the Wild Dog Project is dedicated to breeding and developing packs of the dogs, one day to be released into a network of small fenced-in reserves that will be carefully constructed and managed with the help of donors and partners. After the original adults were brought to Harnas in 1996, four puppies were added when their pack was killed near the border of Bushman territory. Since then five litters have been born, and the dogs have been separated carefully into three camps. The current population stands at over 40.

At Harnas the wild dog enclosures are a favorite stop on the morning tour the guests take. Most visitors go on this tour, as it provides close contact with the fully grown and mostly wild animals on the farm. Etosha, the native Herero guide, drives a safari truck fitted with raised benches that allow tourists to get the best view of the animals. The truck pulls a trailer carrying the foods for the variety of creatures they will encounter. Usually two volunteers accompany the tour to do the feeding as Etosha describes the animals, their habits, and their histories. He is bursting with information, and tourists come away with myriad interesting facts and stories about the animals they have seen.

The first stop on the tour is the largest of the baboon enclosures, holding between 55 and 70 baboons. Etosha explains that in Namibia, farmers regard baboons as problem animals. While they don't kill other animals, they do steal food and wreak havoc through their curiosity, which leads them to break into homes. They also milk the farmers' goats to drink the milk, sometimes so fiercely that they tear the goats' teats. Farmers routinely shoot baboons or trap them. Marieta has worked very hard with local farmers to train them to bring the baboons to her—or if they shoot the baboon and find a baby clinging to the mother, she begs them to bring the infant to Harnas.

Word of mouth isn't enough, though, and getting the information out is now done professionally. Someone from Harnas speaks to farmers about conservation of wildlife at Farmers Association meetings, Harnas advertisements appear in newspapers and magazines—like *Agriforum*—that specializes in reporting on animal life, and for World Animal Day, Marieta used the radio to get out the message of saving wildlife.

In the baboon enclosure, the animals have removed all leaves and small branches from the trees, polishing the wood to a clean, smooth surface. This stripping allows the baboons to sit at the tops of the trees and spot any predator—such as a snake or a leopard—that threatens the troop. A sentry is posted as high as possible, watching for intruders, and if he sees one, he will let out the unmistakable "WAH-hoo!" warning over and over.

While Etosha explains this, the volunteers stand on the moving trailer, throwing a sort of porridge called *mieliepap*, scooping it out with their bare hands and flinging it over the fence to the baboons that come running from all over the enclosure to get their share of breakfast. The *mieliepap*, with the consistency of thick oatmeal, is made from ground corn mixed with whatever leftover food the kitchen has, making a sticky, oozing mess. Nothing at Harnas goes to waste. By the time the four huge buckets have been emptied, the volunteers are covered with it, so Etosha stops briefly to rinse them down, the volunteers laughing through the whole process.

This, too, is part of the Harnas volunteer experience: getting dirty and enjoying it.

The lion enclosures are next, and the amazed visitors start snapping pictures as fast as possible as the lions prowl within 10 feet of the truck, waiting for their morning meal. Sam comes first, then Sarah and Elsa. Sher Khan, the largest male growls and roars and kicks up dirt with his paws behind the fence until his chunk of meat is tossed over the fence. Another enclosure holds brother and sister Sian and Dewi, along with Macho, Simba, and Lerato. Each one receives a large piece of meat, sometimes in the form of a recognizable animal part like a leg or head, and visitors grimace. The scene forces them to realize what we all know but want to repress: all meat begins as an animal, and some animals must die for others to live. Most of us don't have to think about that as we buy our meat in cellophane packages. Here, though, the fact is undeniable, hitting us in our faces—and in our stomachs.

The African wild dogs are next, and Etosha spends longer here, explaining the dire situation of these animals in the wild and teaching the tourists about an animal they may not ever have heard of and certainly have never seen up close. The squealing of the dogs as they wait for their meat is deafening, and as pieces of meat are thrown over the fence, the dogs jump high in the air to catch a piece before another dog gets it. Then they run off into the bushes to eat privately. Slowly the yelps and yips lessen until every wild dog has been fed.

One of the highlights of the tour is the drive through the wild cheetah enclosure, which holds 22 full-grown cats. One of the volunteers is called upon to open the gate, and she is immediately surrounded by eager, hungry cheetahs recognizing that breakfast has arrived. Etosha reminds the volunteer not to turn her back on any cat and to speak to them with authority, meeting their eyes with a fierce stare. She is also reminded to watch out for "the crazy one" with the crossed-eyes and tail tucked between his legs. That one, Etosha reminds the volunteer, is unpredictable. Although cheetahs will back down from this kind of human authority, they will still always be looking for any sign of vulnerability on the human's part.

Once the gate is open, the truck drives through, pulling buckets of meat for the cheetahs. After the gate is closed behind the truck, two volunteers must stand on the trailer and face off with the hungry cats, keeping them from jumping on the trailer and taking the meat early. Adrenaline runs high through the volunteers, for holding off wild cheetahs with just a small stick and an authoritative demeanor is probably one of the most thrilling experiences they will ever have.

The truck rolls along a dirt road toward the center of the enclosure, and the tourists snap photos of the cheetahs sprinting alongside the truck, just a few feet away, surrounding the truck in hopes of gaining an advantage over the next cat. At a clearing, Etosha stops and helps the volunteers toss one chunk of meat to each cheetah. Like the wild dogs, cheetahs yip and growl until they get their share, and once they have their piece of meat, they run off into the bush to eat. Suddenly all is silent. The truck rumbles on. Volunteers sigh with relief, but their eyes are still wide and bright after the spine-tingling encounter.

The tour continues, the volunteers feeding more baboons and some vervet monkeys, and then five leopards housed in separate enclosures. Leopards are solitary in nature, Etosha explains, and only come together for breeding. The stealth hunter, as a leopard is known, never reveals itself until it is too late, announcing its presence by clamping its teeth around your neck. The tourists shiver in response to this news—until Etosha or one of the volunteers reaches into Missy Jo's enclosure to pet the tame-as-a-kitten female. Touching Missy Jo's body is both wonderful and sad. Her coat is thick, warm, and silky-velvet, stretched across the pure muscle flexing beneath, but that is also what is so disheartening. Hunters crave and insensitive women want to wear this amazing pelt, putting the leopard in constant danger in the wild. Harnas has a total of eight leopards—most on the wild side, though, like Honkwe, an animal so aggressive that one time he broke out of his enclosure and killed several female leopards before he could be caught.

After Etosha and the volunteers feed the leopards, they take the guests on a short walk to the caracal enclosure. Caracals are beautiful little cats who stand about 18 inches high and weigh around 40 pounds. Their most recognizable feature is their tall, tufted ears. Despite its relatively small size, a caracal can bring down a small antelope and can jump high into the air and catch a bird in flight. Mostly, a caracal's diet consists of small ground rodents, and if more farmers realized this, this cat wouldn't be so persecuted. Caracals are hard to see in the wild because they live solitary lives and hunt at night. But Harnas has seven beautiful cats that come to the fence when called for feeding. Guests are offered a sight that few see— seven caracals all together, in the daylight and only a few feet away, flying acrobatically through the air as they leap for the chunks of meat thrown their way.

The tour ends with a short drive through another wild dog enclosure; this one housing seven puppies that gallop alongside the truck. And then back to the lapa, where the guests find something cold to drink and even more to talk about, pictures to share, and amazement at the adventure they have been on for the last three hours.

What will tourists take away with them? Is this simply a type of a zoo? Certainly not. Etosha makes sure these people receive an education about the animals of Namibia and the plight each species faces in the wild. These animals are ambassadors for their species, for the land where they live, and for the conservation of wildlife in general. After seeing these beautiful, powerful, and endangered animals competing for meat and loping alongside the truck, these guests will find it harder to participate in hunting, to wear fur, or to ignore the destruction of habitat. Some might even spread the message to others, send their children to volunteer at Harnas, or donate money to help the farm's continued efforts. Ignorance of the plight of African wildlife is impossible after you take one of these tours. Once you know the truth, you can never go back to not knowing.

What will be the future of Harnas? The van der Merwe family has plenty of ideas for what will come. When asked what she wants next for Harnas, Marieta raises her eyebrows and says, "I actually have a book—a dream book—filled with visions. Every month I get together with Jo, Schalk, Melanie, and Frikkie, and we say, 'Now we have enough money for this' or 'We need to spend money on this instead.' Every month we look at our big list of priorities, and as we look we decide what we want to do."

Melanie's next project is one of hope. With Marieta's blessing, she has recently established the Broken Wing Children's Foundation for traumatized and/or abused children, using nature and animals to help these children heal. Having experienced Nico's death and witnessed its traumatic effects in her own family, Melanie knows how children are especially vulnerable to such cataclysmic events as death, abuse, and neglect. She believes that if they live in coexistence with nature, they can see and participate in the healing process. A two- or three-week period in the farm environment at Harnas loving and being loved by animals, Melanie knows, is not enough to heal "broken wings," but she does believe it can offer hope and help put the healing process into motion.

She also plans workshops for children who have Down syndrome, or who are blind or paralyzed, as well as for children with emotional problems from physical or psychological abuse. "Imagine these children seeing and touching tame wild animals!" Melanie says. Marieta has already done this on a smaller scale, taking baby baboons and sometimes Goeters, the tamest of the cheetahs, to retirement homes and to a school for blind children in Windhoek.

Marieta also wants to build a huge baboon enclosure that covers almost four square miles. But money is always short with 400 animals to feed, and the price for the fencing alone on a project like that is astronomical. For the guests, Marieta and Schalk have a dream of putting a house on stilts at every major water hole on the farm. Guests would be driven out to one of these houses and dropped off to spend the night among—but above—the wild animals, watching them and listening to them from a secure balcony. In the morning, they would be picked up before breakfast. And Frikkie's new plans for the volunteer program are also in the works, with special two-week stays reserved for "Mature Volunteers" and for couples who want to come and work together.

Such ideas for evolution and growth of habitats are cognizant of the changes overtaking this once wild continent. From such models others can learn valuable lessons about protecting wildlife while ensuring that the animals live in the way nature intended. Humans do not have to be the ruin of these magnificent creatures. The unique experiment at Harnas will continue to shine a light for many years to come.

Marieta and Zion

{ chapter sixteen }

THE SOUL *of a* LION

When a man wants to murder a tiger he calls it sport;
when the tiger wants to murder him he calls it ferocity.

—George Bernard Shaw

IN MAY 1997, THE SOUTH AFRICAN EQUIVALENT OF ABC'S *20/20* or NBC's *Dateline*, a program called *Carte Blanche*, conducted an undercover investigation of "canned lion hunting." Canned hunting is usually defined as shooting an animal that has been "tranquilized or artificially lured by sound, scent, visual stimuli, feeding, bait, or any other method," according to Dr. Pieter Botha of South Africa's Environmental Affairs and Tourism. In other words, a tame lion who has never lived in the wild and doesn't fear people is paraded out in front of a hunter at close range, and the hunter shoots it, taking the head or pelt home and telling his friends about his "dangerous" hunting experience in Africa.

The result of the investigation proved horrifying, as demonstrated by the central incident, narrated by witness Bruce Hamilton:

The lioness had three cubs. We took her out of the camp that morning into a hundred hectare enclosure, which was not legal. And she was still running up and down the fence. She wouldn't leave her cubs, even though bait was used to try and lure her away from the fence so that the hunter wouldn't see the fences and be caught up in the illusion. Even though she wouldn't leave the fences, he still shot her. He could see the cubs on the other side of the fence, but that didn't bother [him]. Even when the lioness was skinned and the milk was pouring out of her teats, it didn't bother the hunter that she was still producing milk for those cubs, and now they didn't have a mother.

The show so outraged the public that then minister for environmental affairs and tourism Z. Pallo Jordan placed a moratorium on the breeding of lions in captivity. Unfortunately, this moratorium was "voluntary," and with over 2,500 lions in breeding facilities throughout the country, and over 400 lions killed per year this way—at a price of between $13,000 and $30,000 per animal—very little has changed in the years since the investigation. Lions are hardly the only victims. In the North West province, in one of the most exclusive hunting lodges in South Africa, investigators have found Bengal tigers, black panthers, brown bears, pumas, leopards, jaguars, chimpanzees, cheetahs, and African wild dogs, all waiting to be killed by trophy hunters during canned hunts.

Who is spending all this money to shoot tame, helpless animals? Some of the hunters are German and Australian, but the vast majority of the people driving the demand are Americans. Piet Warren, a member of the Limpopo Carnivore Association, is a businessman and farmer who has been breeding lions and other animals for hunting for years. He says some Americans come to Africa and want to kill ten animals in one week. Why would he voluntarily give up this thriving business when there is "somebody who's standing in the street in America shouting, 'Piet Warren I want to give you this fat cheque for that lion which I want to hunt orderly in a nice sizeable place . . . because I'm [scared] of the bloody thing!'"

Consider this horrifying situation in contrast to the lions at Harnas, where they are treated as individuals. Zion and Trust were raised from birth at the farm—and assigned a veteran yellow lab, Tara, as their "dog mother." Tara has raised many lions and has the scars to prove it. She is a wise, veteran lion mother, despite her advancing age and expanding belly. She takes no flak from these lion brothers, both offspring of father Hemingway and mother Simba, but from different litters born five months apart—Zion in December 2005 and Trust in May 2006. (The gestation period for lions is only 110 days.) Volunteers worked closely with these lions from the time they were cubs, as they do with all of the baby animals. Frikkie is always careful about introducing the lions to the new volunteers—especially once the "babies" grow to be 200-pounders who still think they are cubs. Frikkie tells the volunteers to get used to the lions and to let the lions get used to them, slowly. Until everyone is comfortable, a volunteer cannot enter the lion enclosure alone.

Veteran volunteers often coach the new ones on how to approach Zion and Trust, explaining the rules about never turning your back on them, never crouching down, never screaming or running—and warning new people that Trust is inappropriately named. He loves to sneak around and jump on your back in play. That was fine when he was a 40-pound cub, but once he grew to be more than five times that weight, he could easily knock anyone to the ground if he chose to—which he occasionally does, out of fun, not aggression.

Volunteers do the feeding of Harnas lion cubs, they clean their enclosure, they take them for walks out in the bush, and they create toys and games to keep them occupied. Some brave the cool nights to sleep with the lions in their enclosure or out in the bush. But sometimes volunteers pay the price, like Rachel, a volunteer who woke up to find Trust methodically chewing through a chunk of her hair. Through their own experiences, volunteers learn to read animals' faces and to see when they need attention or want to be left alone. They learn to respect these soulful creatures that are growing rarer in number. Each has a personality, and along with that, most volunteers who have worked and played with them agree that each has a soul.

How many people get to "play" with lions? And how could someone, seeing how tame and playful these creatures can be if they are handled carefully, want to shoot and kill one just to hang its head on a wall? It's shocking that this question needs to be asked, and yet canned lion hunting is growing in popularity. Habits die hard, especially when there are easy profits to be made. In fact, on the Internet it is now possible to hunt by remote control. For anyone who has the money and a computer, it is possible to shoot a lion from the comfort of one's living room.

Marieta's son Schalk has lived with lions his whole life. He wrestles with them and plays soccer with them. When asked about this, he shrugs his shoulders and says, "Some people have dogs in their houses. It's just the same thing for me, except it's lions. I never realized what I had until outsiders showed me how special it was."

Now that more time has passed and Zion and Trust have grown beyond the cub stage into adult males weighing more than 500 pounds each, they have become more aggressive and territorial, and volunteers aren't allowed to tangle with them. But Schalk still goes out to their large new enclosure and spends time with them as often as he can. He smiles, explaining, "I'm more like a playmate with them because we got them so young. Zion is very gentle and likes to be around people. It's Trust who is the naughty boy. There might come a day when something will go wrong. Usually what people do is blame the animal, but I hope I won't do that. I have to respect them." When he is asked how he will feel when he knows he can no longer walk into their enclosure and play with them, he takes a minute, swallows, and admits, "It will be hard."

Schalk describes Zion as extremely attached to him, and, like a loyal dog, Zion wants to be with his owner all the time. He even feels abandoned when turned loose in the bush to run free. When Zion was only a year old, Schalk took him out with a group of tourists. The plan was to drop Zion off out in the bush and then drive the tourists to the water hole. Schalk was confident that Zion would have a good time exploring and then find his way to meet them at the destination. But after waiting two hours for the lion, Schalk thought he'd overestimated Zion's tracking skills, so he drove back to the place he'd dropped off the cub. There he was, right where Schalk had left him, looking morose and neglected. "Abandonment issues," a psychologist might diagnose, but when Zion was released, he no doubt trusted that Schalk would return for him. He perhaps was just upset that the wait was longer than he thought it should be. Despite his massive size, Zion is definitely a "Daddy's Boy," laughs Schalk.

Trust may no longer be predictable enough to take on bush walks, but under Schalk's guidance, Zion remains daunting but gentle. Schalk calls Zion "a special boy" who, despite his physical and sexual maturity, can still be disciplined enough to walk beside humans and interact with them on a limited basis without attacking. That's one reason Zion and Trust are set to join the Lifeline Project in the near future. In case of an animal's illness or injury, people wanting to help will need to be able to approach a large predator without darting it.

Zion will soon be the largest and most powerful lion at Harnas, surpassing even Sher Khan as the king. Sher Khan has long been untouchable by any human—even though he was raised from a cub. He kicks his back feet in an act of defiance if anyone comes too close to his fence, roaring his displeasure at the encroachment on his territory. Most male animals like lions, leopards, baboons, and warthogs become aggressive and territorial once they mature sexually, even when they are raised with humans. Female animals remain relatively more approachable. This is why Missy Jo, Elsa, Pride, and Cleopatra are still tame enough to interact with humans, but Sher Khan, Macho, and Patcha are not. This knowledge of individual animals can make the difference between a joyful encounter and a tragedy.

Amid the constant collisions of animals and humans, Harnas will never lack for more animals. People will, unfortunately, always kill "problem" animals, sometimes finding the babies they have orphaned. People will continue to choose exotic animals as pets, usually at some point realizing they have made a mistake and so look for a place to get rid of the animal. Animals will continue to be injured or abandoned, and they will continue to get sick. The lucky ones will find their way to Harnas and to Marieta van der Merwe. And she will take them in—all of them. She will heal them, feed them, raise them, name them, and love them. And if she can, she will release them to become free again.

Recently, Harnas has had some interesting new arrivals—six new cheetah cubs, some baby baboons, and in September 2008, Simba gave birth to four male cubs, and another litter of four lion cubs—two males and two females—arrived in February 2009. As a rule now, when cubs are discovered, they are removed from the mother's enclosure, as many lion cubs end up dead at the hands—or paws—of other lions, usually either the father or another female. Both Zion and Trust, for example, had siblings killed before Marieta could get to them. So the two recent litters of four cubs each were taken out and are being raised near the family house. The journey begins again, and Schalk now has eight new soccer opponents to wrangle with.

Such wondrous creatures as lions are incredibly powerful, easily outmatching us humans physically, but we have used our intellect and technology to gain dominance over them. This power doesn't have to be negative, because we can help and learn from them, but humans, unfortunately, too often use their power to exploit wild creatures rather than protect them, behaving as tyrants of the planet rather than its guardians.

Despite this, the smiles on the faces of the van der Merwe family couldn't be more joyful when a new animal arrives, even knowing that for every animal saved, many more are lost. Even though they live here every day, they still understand the magic of Harnas. Jo claims, "it's a miracle place," and Schalk describes it as "like a dream—but with a purpose." Frikkie knows "All the people on Harnas are living with a passion. It's not about 'me.' It's about the animals."

As the animals keep coming, so do volunteers and ecotourists. Again and again they make Harnas a second home. Do they just want to see the animals or is the impulse more deeply rooted? Interview after interview of people who live, work, and volunteer at Harnas reveal the same word over and over: *healing*. True, animals are healed, but people who visit and help these animals benefit themselves. Frikkie says, "The closer you get to animals, the more your feelings open. You can laugh, you can cry. You learn about yourself again." Jo puts it another way in an interview with Animal Planet: Harnas is "a miracle place. The fire in people's hearts [that they have when they are young] starts to dim down and die out. When you come to Africa, [that place in your heart] starts to feel a little bit of life again, and you start to feel emotions, and your blood bursts with life. And that's partially why people say you get the 'fever,' the 'African fever.' You keep having to come back."

The center of all of this activity is Marieta, the creator and protector of Harnas, who has the soul of a lion: fierce in her determination and protection of those she loves, but also nurturing and gentle when needed, considering the welfare of her "pride" first and foremost. This was not her original plan in life, but the path opened up and she gladly took the journey. Harnas began with this one woman rescuing one animal from the side of the road, and from that time forward, Marieta's desire to help has never wavered. She tells the world, "If you love animals, come to Harnas. If you have a passion for animals, this is the place to be."

Marieta feels chosen for this important role in the world. "Harnas. Mmmm," she considers. "I'd only like to say if I could make my choice over again on earth, I would want to be here on Harnas. For me it's the best place on earth. I'm so happy that the choice was *me* to be *here* with the animals."

ACKNOWLEDGMENTS

FIRST AND FOREMOST, I'D LIKE TO THANK MARIETA AND the rest of the van der Merwe family for opening their lives to me and allowing me to listen to their fascinating stories. It's not easy to tell both the good and the bad, the pleasant and the painful, but they did so with honesty, and they made me feel a part of the family while doing it. Thanks, too, to the many volunteers who offered me their stories and pictures. Research doesn't get more fun than this.

I'm grateful to North Carolina State University and especially the College of Humanities and Social Sciences for giving me the time to write this book and the Scholarly Project Award, which helped fund some of the many research trips I took to Namibia. The Department of English has also been very supportive as I've tackled this new kind of writing project, and I'm indebted to my fine colleagues.

Thanks to journalist Britt Collins, who believed in my project and so introduced me to her agent, Ellen Levine. And thanks to tenacious Ellen for finding such a perfect publisher as National Geographic, with its tremendous staff headed by Barbara Brownell Grogan. I couldn't have picked a better home for this story and for me.

Kevin Mulroy, editor extraordinaire, took my raw work and helped me make it better—and never seemed to lose patience. I also appreciate an early reader and editor, Linda Whitney Hobson, who gave insights and suggestions to improve every page.

The book contains much medical knowledge that I didn't have when I started, and so I thank Dr. Graham Snyder for answers to everything from heart attacks to baboon births. The information on Crimean-Congo Hemorrhagic Fever came from several doctors and researchers who helped me simplify a complicated disease: Bob Swanepoel, Onder Ergonul, Dan Shapiro, and Jack Woodall.

On a personal level, I couldn't have written the book without my Girls Night Out friends: Marianne Gingher, Susan Irons, and Jill McCorkle. Long-distance shoulders came from Joanne Greifer, Nancy Houston Jones, Gay Hanby, and especially Amy Leander. And the girls in my neighborhood from the next generation that will be given the task of saving the world, big thanks: Claire, Kirstin, Hanna, Laney, and Carolyn.

Finally, to my brother Randall for keeping me relatively sane, and to Fred (along with my dogs, Henry and Milo), who supports me and keeps me stable. This wouldn't have happened without you, boys.

INDEX

WHAT CAN YOU DO TO HELP?

Go to www.harnas.org

Visit • Donate • Volunteer • Adopt an animal
Sponsor a Bushman child • Support a project